Mixing a Musical

Mixing a Musical: Broadway Theatrical Sound Techniques, Second Edition pulls the curtain back on one of the least understood careers in live theater: the role and responsibilities of the sound technician.

This comprehensive book encompasses every position from shop crew labor to assistant designer to sound board operator and everything in between. Written in a clear and easy to read style and illustrated with real-world examples of personal experience and professional interviews, Slaton shows you how to mix live theater shows from the basics of equipment and set ups, using sound levels to create atmosphere, emotion, and tension to ensure a first-rate performance every time.

This new edition gives special attention to mixing techniques and practices, and special features of the book include interviews with some of today's most successful mixers and designers.

Shannon Slaton is a sound designer based in New York. He has designed the tours *Elf, Kiss Me Kate, Noise/Funk, The Full Monty, Contact, Tap Dogs, Hairspray, The Producers, Sweeney Todd, The Wizard of Oz, The Drowsy Chaperone, Sound of Music, Bullets over Broadway, Finding Neverland, Memphis, Amazing Grace, and The Wedding Singer*. His Broadway work mixing includes *Man of La Mancha, Bombay Dreams, A Christmas Carol, Sweet Charity, Jersey Boys, Dirty Rotten Scoundrels, The Drowsy Chaperone, Spring Awakening, Fela!, Anything Goes, Holiday Inn, Annie, Cabaret, Springsteen on Broadway, and Legally Blonde*. On Broadway he designed *The Illusionists* at the Palace. Off-Broadway he assisted on *Hurly Burly* and on Broadway he assisted on *Steel Magnolias* and was the Associate on *Barefoot in the Park, Blackbird, An Act of God, Present Laughter, Meteor Shower*, and was the Advance Sound on *Wicked*. Regional designs include shows at George Street Playhouse, Casa Manana, Maine State Music Theater, and North Carolina Theater.

Mixing a Musical
Broadway Theatrical Sound Techniques

Second Edition

Shannon Slaton

Routledge
Taylor & Francis Group

LONDON AND NEW YORK

First edition published by Routledge 2011

Second edition published 2019 by Routledge
2 Park Square, Milton Park, Abingdon, Oxon, OX14 4RN
52 Vanderbilt Avenue, New York, NY 10017

Routledge is an imprint of the Taylor & Francis Group, an informa business

Library of Congress Cataloging-in-Publication Data
Names: Slaton, Shannon, author.
Title: Mixing a musical : Broadway theatrical sound techniques / Shannon Slaton.
Description: Second edition. | New York, NY : Routledge, 2018. | Includes index. Identifiers:
LCCN 2017059500 | ISBN 9780815367482 (hardback : alk. paper) |
    ISBN 9781138491441 (pbk. : alk. paper) | ISBN 9781351033060 (ebook)
Subjects: LCSH: Theaters—Electronic sound control. | Auditoriums—Electronic sound
    control. | Sound—Recording and reproducing.
Classification: LCC TK7881.4 .S57 2018 | DDC 792.02/4—dc23
LC record available at https://lccn.loc.gov/2017059500

ISBN 13: 978-1-138-49144-1 (pbk)
ISBN 13: 978-0-815-36748-2 (hbk)

Typeset in Utopia
by Apex CoVantage, LLC

To my wife, Mollie, thank you for all of your support.

To my son, Parker, you make me believe in superheroes.

To my daughter, Lizzie, yes, you are a princess with magical powers.

Also, to three of the best mixers I have ever known, Jordan Pankin, Bob Biasetti, and Francis Elers. I have learned most of what I know from these guys.

To Patrick Pummill for being a great friend and one of my favorite people to talk shop with.

Finally, to Scott Armstrong for being my favorite person to build a show and debate with. You are missed by me and many others in the sound community.

CONTENTS

INTRODUCTION—WHY THE BLEEP AM I WRITING THIS?

As I begin what I know will be a laborious and grueling task, this is the main question I have. Why am I doing this? Why am I about to put myself through this? Hours and hours of sitting alone and trying to work myself into the mood of writing. Bleh! I remember hearing a story about Tom Waits (if you don't know him, then stop reading and go listen to "Frank's Wild Years" before reading any further. I will wait.). Anyway, I heard a story about how Tom Waits has a muse and he talks to the muse all the time while writing and recording an album. Story goes, Tom is trying to record one more song for an album and they do take after take and it's just not working. Tom, or Mr. Waits—both seem inappropriate. I don't know him, so I shouldn't call him Tom, but I feel like I know him because he has lived in my ears for decades, so Mr. Waits seems too formal. Anyway, Tommy is talking to his muse about this song and he's really frustrated because it's just not working, and he says something to the effect of, "Alright, this is your last chance. The bags are packed and the car is loaded. Now either you get in the car with the rest of the songs or we are leaving you behind." Well, let me be the first to tell you that a muse for writing about technical theater and sound mixing does not exist. I have searched, but there is no invisible creative friend to prod you along and give you inspiration. It's not fun. It's not exciting. The only reason I am writing this is because I have something to say about this craft and, bleep it, I feel compelled to write it.

When I was asked to write the first edition, I had no clue there would be a second edition. I was very excited at the thought of putting my thoughts and experiences down on paper to share with people and hopefully learn 'em something. I wanted to write the book because I loved theatrical sound mixing and wanted to share the experience of having a career in theatrical sound as a mixer. I wanted to show people it was an art form as well as a technical challenge. I wanted to show people that it was an exciting career path—but even then, I knew I was taking a big risk by writing it. I mean, seriously, who am I? I'm just a normal old sound mixer. I'm not the best. I'm not the first. In fact, in the world of mixers, I would put myself in the third generation of mixers, and I only say that because most of the mixers I learned from had learned their craft from the first generation of mixers. This craft was so young when I started that I could say I was learning from a guy who learned from a guy who was among the first people to put a mic on an actor. Now he had some stories to tell and a lot to teach. So, who am I? (Cue the *Les Miz* music "24601.")

When I wrote the book, I told very few people what I was doing. I was worried about getting made fun of. Let's face it. Theater is a tough business full of people who like to rib each other. I am as guilty as the next stagehand. I mean, if you are not careful you will be marked for life for the simplest mistake. If you fumble tying a clove hitch one day, you might end up being called "Hitch" for the next 20 years. I know a guy who people call "Bowline" because he had a meltdown onstage when someone tied a bowline on something that did not in any way require a bowline. I figure he has another five-year sentence on that nickname, but he might get paroled for good behavior in two to three. So, suffice it to say, I was a bit nervous for the Broadway community to learn about my secret. I knew there would be ribbing. I knew there would be snickering behind my back. I knew there would be nicknames. In fact, one person whom I respect very much sent me an email when they found out. It very tersely and in not so many words said, "Who the bleep do you think you are?" I still respect this person very much and hopefully my sentence is not life for this. I wish this person could've seen that what I was writing was to honor and pass on the passion this person had helped instill in me. But I had expected that reaction by some.

I think, by far, my favorite ribbing happened where lots of good ribbing happens. In a bar. Now, when I tell you about this, I want it to be a lesson that you need a thick skin in this business; if you can't laugh about things, then it is going to be rough for you. Broadway sound people are a close-knit group of cutthroat pirates, and we like to get together. We all know each other, and we enjoy hanging out together. I don't see the same with other disciplines, but maybe it happens, and I just don't know. For decades on Broadway, sound people will choose a night and a bar and will meet and laugh and talk about whatever we are doing. I can tell you it is almost always on a Thursday, but I can't tell you where.

One night, about a year after the book came out, I was talking to a designer/mixer I knew. We have known each other for probably 15 years, but we have never actually worked together. I basically only see and talk to her at these get-togethers. But we have always been friendly. Out of the blue and four pints in, she brings up my book. I never bring up my book at these events. In fact, I don't mention my book within a ten-block radius of Times Square, just to keep from getting a nickname. But she brought it up and said she hadn't read it. I explained that I wrote it to get young people excited about a career in sound. She said, and this is almost verbatim, "I think that is great. And you know (hiccup), I feel like (hiccup) a book like that needed to be (hiccup) written. I just wish that it had been written by someone with (hiccup) more experience. You know." I replied as only I could. "Yeah. I know. I agree." That was what I was afraid of when I wrote the book. I was never trying to put myself forward as an expert. I learned from experts. I developed a passion for the craft. I felt more like a conduit passing on what I had learned. But to be fair, my resume is not too shabby.

By the way, I still talk to her and consider her a friend, and I totally respect her opinion. Thick skin.

So why am I willing to do this again, open up these wounds and take a chance of getting pegged with some god-awful nickname? There are a couple of reasons. First, I don't think I finished. Writing a book takes a lot of time and mental patience. It just roams around your brain all the time. It is exhausting. I am proud of the book I wrote, but I feel like I left some things out. In a way, I ran out of steam. I didn't have the stamina to do any more. Over the last six years, I have had time to think over what I wrote and what I didn't. Every once in a while, I will be at a gig and just think, "I wish I had put this in the book." In my career, I straddled the world between mixer and designer for 20 years. Recently,

I have moved away from mixing and more towards solely designing, and I have worked with a lot of mixers at different levels. There have been times when a young mixer would ask me a question and I would jokingly say, "Read Chapter 5 of my book. The answer is in there." But there are lots of other times when a question is asked and I realize I didn't put this in the book. I'm hoping to fill some of those holes.

Another reason I am writing this edition is because of a fair criticism I have received from some people. I have been told that I didn't spend enough time writing about actual mixing. I think this is probably true, and I plan to go on a deeper dive in this edition, but this also points out a failure in the first edition that I hope to rectify in this edition. One of my main points in the first book is that being a mixer is a lot more than moving faders up and down and pressing "GO." It is about the minutia and the politics. It's like making a cake. Mixing is like icing a cake. It's fun and creative. It's what everyone sees as being a mixer. But you don't get to that point without first developing a recipe and buying the ingredients. Then you have to assemble the ingredients, and you better know that you mix your wet, then your dry, and then together. You have to know how to properly grease a pan or it won't slide out. Then you have to put the layers together and make the icing. Then you get to ice the cake. I am going to double-down in this edition to make it clear that you have to know it all to be a good mixer, as well as go deeper into the mechanics of mixing.

An example of this is the show I am designing right now. I am sitting in the Brunswick Tavern in scenic Maine waiting for my breakfast before heading off to a 10 out of 12. Just in case you haven't heard that expression before, it means we will be working in the theater with the actors for 12 hours with two meal breaks. Of course, I will be there two hours before the cast and an hour after, so it is really a 10 out of 15 for me and the rest of the creatives and crew. The mixer at the theater is Nate Dickson. He's a great guy and really on top of it. What makes him a great mixer, to me, is not how well he moves the faders. What makes him really good at what he does is that he had a well-thought out plan and he executed it. Com and video were up and running with no problems. The pit was set up and tested a day before the band moved in. His crew (hi, Julie, Rachel, and Connor) have busily prepped mics for the actors. It all runs like a smooth process, and because of that Nate and I have been able to spend time with clear heads developing a strategy for mixing the show and programming the board. Heck, we've even had time to laugh and joke with very little stress, and that should be the goal. By the time you sit down to mix, it should be a joy, and the only way that can happen is if all the other bleep has been taken care of.

There is another reason I am writing this. I have found over the years that musical theater is not always part of theatrical sound design education. I am not saying that it isn't taught anywhere, but it is rare. There are some very good programs out there to learn how to design and mix musical theater. But I know there are lots of places where it is not a major part of the curriculum. I remember years ago asking a professor of sound design why his program didn't offer any courses in musical theater and he said it just wasn't the focus of the program. They specialized in plays. In preparing to write this book, I asked some people for thoughts and opinions and was told by one person that "very few programs (undergrad or graduate) provide a specific course in sound for musical theater." How is that possible? Why would a sound design program ignore a very important job opportunity in theater? I am currently designing a show in a theater that only does musicals. And this is one of three theaters I have designed for that only does musicals, and without much effort I can think of half a dozen other regional theaters that only do musicals.

There are currently 40 Broadway theaters, and I have climbed into the ceiling of approximately 26 of them. Currently there are 30 shows playing on Broadway, and of those 30 shows, 25 of the shows are musicals and only five are plays. There are currently 24 Broadway tours traveling the country, and of those 24 shows, only one is a play. If you look at USA829 contract minimums, the fee and weekly for designing a musical is almost 70% more than for a play. Even at regional theaters, most theaters have at least one musical a year. It's time we stop ignoring musicals in theatrical sound education.

I can understand some of the reasons musical theater is hard in the educational world. In order to teach it you need a program that performs musicals. How can you really learn about mixing and designing musicals unless you get to do it? My college only did two musicals in the four years I was there. I mixed both musicals, but it definitely wasn't enough to build a complete learning experience around. But there are ways to learn to mix musicals without actually having live productions, and I plan to explain that later in the book.

There is another important thing you need in order to teach sound for musicals. You need a decent-sized theater and a proper sound system. My college was lacking the latter. We had a decent-sized theater, but there was no sound system. The two musicals we did were supported by a local sound hire company that mainly did rock and roll and corporate gigs. But, again, there are ways to learn this without having a large theater and an amazing sound system. One of the people I talked to suggested I go into depth on how to EQ a mic and tune a system. As I stated in the first edition, that is not the purpose of this book. If you want to learn about EQ and tuning, you should read Bob McCarthy or take a seminar with the people from Smaart. I have mixed shows where I never touched the EQ unless there was a cover in the show. I have worked with designers who didn't want me to touch the EQ, and I can't think of a single designer who would want me to touch the system EQ. But there are concepts that you need to understand as a mixer that I will go over in this book, and there are exercises you can do to learn about sound for a musical without having a full system in place.

It astonishes me sometimes how overlooked sound in musical theater can be. I have several friends at the World Stage Design Conference in Taipei right now. About 12 years ago, for the very first World Stage Design Conference, I submitted a design I did of the national tour of *The Full Monty*. I filled out the application. I made a packet that included signal flow drawings and speaker tower drawings. I even made a video with pictures of the production and the racks and equipment, and I set it to music from the show so you could hear the mix of the show. I was rejected by the conference because they didn't see the value in rack drawings and signal flow drawings. They also said that I could only use original music. I argued that if you require only the use of original music, then the World Stage Design Conference would never be able to showcase musical theater sound designers. Here we are 12 years later, and they still do not allow musical theater sound designers to be part of the conference unless, of course, that designer also wrote the musical. In my opinion, it seems that what they consider to be sound design is actually musical composition. As someone who makes subtle decisions about speaker placement and reverb and level to help push the story forward, it is frustrating to find that work is not considered sound design as much as writing a song is.

Another crucial element you need—actually, the most important element you need—is teachers who are able to teach sound. From my many visits to USITT, my experience is that this is not a problem for most programs out there. I even know some retired Broadway musical mixers who have started teaching. Now, that is awesome. But that is not always the case. Sound is very specialized and can appear to be voodoo to lots of people. I have a friend whose sound classes were taught by the lighting professor, who tried to explain how you have to set the "intensity" of the fader correctly. My friend was like, "Yeah, I don't think it works that way." It is not uncommon for lighting people to teach sound. After all, us sound people are just assistant electricians in the eyes of the union. And the reality is that our worlds are not that much different. They just seem to be. I have worked with several lighting teachers to help them understand sound so that they could incorporate more into their classes. I can only hope this book helps a teacher in that situation. I have a lot of respect for the lighting teachers who are also tasked with teaching sound. That is a lot of knowledge to keep rattling around in the old noggin. I can't imagine doing it. The longer I am in this business, the less I seem to know about anything except sound.

But, most importantly, you need to see the value in sound for musical theater. Mixing and designing sound for a musical is a skill and a craft and an art form. It is easy to overlook it as just technical. There are people who honestly think sound people are ruining musical theater. Once a year at Town Hall in New York, an organization does a fundraiser show called *Broadway Unplugged*. Here is the description of the show this year: "Broadway Unplugged returns: great show tunes, great Broadway stars, great (big) voices, and

NO microphones . . . just the pure human voice, the way it used to be on Broadway!" I find it a bit hurtful that I care so much about musicals and work so hard on shows and then the community has a whole show about how much better life is without my field's contribution. But it's a sentiment that is out there. And there are people who believe that what we are doing is just amplifying other people's art, whether it is a singer or a musician. Some people can't appreciate that we are part of what makes that art work. We have to respect the value of what sound mixers add to a musical. Theatrical musical mixers and designers are not composers. We are technicians and tinkerers, and we are musicians playing a very unique instrument made up of dozens of actors and musicians. We have to respect the art form of mixing theater.

But the day is coming that I have been predicted for a few years. More and more plays are mic'ing actors the way we do in musicals and are expecting plays to be mixed similar to a musical. I was recently Associate Sound Designer for *Present Laughter* at the St. James. We started out with the show being completely area mic'ed. I designed and installed a reinforcement system that could've handled a decent-sized musical, and I installed 24 microphones all over the set. I warned the designer, Fitz Patton, and the director and several other people that I didn't think area mics would work for the show, but we forged ahead on the road of bad intentions our area mic'ing forefathers had laid. For the record, I hate area mic'ing. It plain and simple does not sound good. Ever. You end up amplifying distance. The louder you turn up the mic, the further away the actors sound.

We tried and tried to make the area mics work, but this was a show with lots of fast dialogue and quick one-liners. It was not going well. You could hear the actors, but it wasn't enjoyable. Now, I know all the classic tropes. "Actors aren't trained to project anymore." "Audiences aren't trained to listen anymore." These memes go on and on blaming everything except the reality that the world has changed. No one wants to see a play where Ethel Merman screams lines to the back of the house. No one wants to block plays where the actors move from one area mic to another. No one wants to watch a show that has cool projection and rotating sets and flying actors and full surround sound rainstorm scenes and huge scene change music and an acoustic voice that you have to strain to hear. No one! So after weeks of trying, we finally decided to mic the actors. I had to put mics on the cast in the morning, and without a run I had to mix the show that night in front of 1,500 people and some critics. All I can say was the difference between the audience reaction before mics and after mics was staggering. We went from tepid applause to raucous show-stopping applause.

The world is changing. The equipment is more sophisticated. And I am writing this second edition because it is more important than ever that we educate and pass on knowledge and grow this art/technical form. It also can't hurt to have a book with some insight from a person who has spent a majority of his waking hours for the past 25 years working in musical theater.

The final reason I am writing this is because I have more bleep to say!

FOREWORD

As a sound designer, one can spend months, sometimes years, planning out a sound system that will complement the show, satisfy paying customers in every seat, fold into the design elements of sets, lights, and costumes, fit within the budget, and, perhaps most importantly, justify the confidence that the director, composer, and producers have entrusted in you to make their show aurally soar.

The process for me begins with the reading of the script. You learn what kind of show this one is. Perhaps it's a rock musical with a small band onstage; perhaps it's an old-school musical with a large cast and large orchestra seated in the pit. Maybe it's British farce with lots of quickly delivered laugh lines. Next, I visit the venue and calculate mathematically what is needed to service this particular show in this particular theater at the appropriate level of volume or energy. Then begins the tedious weeks of drafting and paperwork required to present a specification that will be bid on by vendors eager to secure a show that may or may not bring financial reward. The phone starts to ring. It's a producer saying you're over budget. It's a vendor asking you to substitute what you believe to be the perfect choice of loudspeaker for one that will cost less and help secure the bid. It's a set designer calling to inform you that there's been a change and you'll need to come to their office and work together and re-trim the height of all your calculations. (Just kidding—that call never comes. You find out during load-in and you have to quickly adjust to the unmentioned change. Ah, collaboration.)

I'll spare you, dear reader, the following weeks from bid award to installation, but just know that, for me, they are often filled with anxiety and second guessing. Fast forward to the first day of rehearsal in the theater. The system has been prepped by the rental shop following squiggly lines you've drawn on paper. The stagehands have hung the speakers to your specification and focus. The cast comes onstage after weeks rehearsing off-site. The stage manager calls "places," the house lights dim, and we're about to begin. You take a deep breath and, like a parent sending their child off to school for the first time, you entrust all the expertise of your design to the ears and fingers of your sound mixer. I can think of no other design discipline that so heavily relies on the talent and focus of another person for its success. Like the actor to the playwright, the relationship of mixer to designer is one of utter dependence. As the actor must take the written words and give them depth and a personality, so too must the sound mixer breathe life into a sound design.

As time goes by and theater technology grows more and more sophisticated, I'm very happy to employ these technical advances to my benefit. Sound systems have become more efficient, and I think this is a boon to all who produce, design, and mix. Ultimately, it's the audience who should reap the benefit. It's what we're here for—to make a quality product that can be enjoyed by folks who pay for a ticket and bring their dates and their families to sit among a crowd of strangers, transported for a few hours in ways that, hopefully, will last in the patron's mind and spirit for a long time. Sitting there among the crowd is the sound mixer, literally sharing the experience in real time. This mixer, board op, or engineer, if you prefer, has all the tools of technology at their fingertips. They understand that with a press of a button the cue will change, bringing with that touch an infinite number of possibilities that will affect the show. Reverbs can expand, microphones re-allocate themselves, dogs bark offstage, EQ can soften. All of these and many more reactions to the technology we've harnessed can occur, but none of it matters—and this is my favorite part—unless the human touch of the sound mixer is present. The world spins modern wonders, but still, a good sound mixer must use their human senses of touch, hearing, and the ineffable quality of taste to achieve a proper sounding show.

As a long-time sound person who began by mixing Off-Broadway shows that led to touring shows that led to Broadway shows, I made the transition some 25 years ago to full-time designer. Since then, I have been blessed with a design career that has allowed me to be part of some great works of American theater. My association with directors and composers and musicians has filled me with a great sense of worth, but I have to say that I often envy my operator. The quality of a sound design is open to endless debate—a good mix is a tangible.

There is a moment at the end of a show when the last lyric has been sung and the final note has been played. Perhaps the ending is plaintive and sad, perhaps bombastic and joyous, but the sound mixer makes their last moves and slides the audio into the blackout. The mixer has an inert moment as the audience erupts into applause. The actors take their due adulation and, in that static moment of the mixer's stillness before preparing for the last bit of music that often accompanies the crowd's departure from the theater, I believe that there's a piece of the applause that belongs to the sound operator who, for the last three hours, has quietly performed as electronic conductor/manipulator. As talent support we don't expect the spotlight, but for the efforts of a mixer who, alone and in the dark, has participated in the live event and contributed fully to the evening, I believe there exists an unheard ovation. That is why I miss mixing.

In the following pages of this book Shannon will take you along for the ride an operator goes through towards the ultimate goal of a pristine mix. It has been my pleasure to have worked with Shannon multiple times over the years, and I am confident that he will translate his practical experiences and those of his colleagues into a relatable and informative book, one that will hopefully finds its way into the hands of people who'll use this knowledge to improve the sound of their local community, college, or regional theater. Who knows? Perhaps a future Broadway mixer will be born. Good luck.

—Brian Ronan

PREFACE

Before I get started with this book, I feel it is important to answer some key questions like . . . Who am I? Why am I writing this book? Who is it for? How is this book laid out? And . . . if a speaker feeds back in the woods, will anyone hear it? So let me start with the first question. Very simply, I am a sound person. I will do and have done almost any job in the field of theatrical sound. I also consider myself a stagehand. I am willing to work anywhere I am needed in the theater, but I probably shouldn't be allowed to wield a jigsaw or repair a wiggle light. I started dabbling in theater in high school back in the mid-1980s. I was a spiky-haired little new-wave punk looking for a place to fit in, and for some reason the theater department drew me in. My first venture into theater was as the newspaper boy in Thornton Wilder's *Our Town*. Ah, those were the days. At that time, I considered myself an actor, but I was one of the few who hung around to build the set and hang and focus the lights. I remember doing musicals back then with no amplification. Nowadays there are high schools running 30 wireless on a musical. It just blows my mind how much things have changed in 25 years.

After high school I went to college and studied English and Theater. What a money-making combo that is! I considered myself an all-around theater person and enjoyed acting, directing, writing, and teching shows. Right after high school and during my first year of college, I worked in lots of little black box theaters around town, and more often than not I was the electrician/light board op. There were also times when I would stage manage as well as build sets. Basically, I would do whatever I needed to in order to get an invitation to the opening night party. Then I became more involved in my college program and did a little of everything. I think wardrobe crew was my least favorite, but other than that I enjoyed what I was doing. I am a big believer in a liberal arts education, and I think the field of sound is a perfect example of that style of education, which encourages you to learn a broad range while you focus on your goal and gives you a rich palette of knowledge to use in your career path. I hadn't really found my niche in theater; I just knew I wanted to work in theater.

Sound was barely a part of theater back then. I remember seeing *The Phantom of the Opera* and *Les Misérables* on Broadway in 1987, and I know those shows were amplified, but in my little world in Texas there was not a lot of sound in theater. Neither my high school, nor my college, nor my local community theater, nor my neighborhood avant-garde theater owned a single wireless microphone. In fact, other than pre-show and some scene change music, I don't even remember any recorded sound cues in anything I worked on until the last couple of years in college. During my junior year in college we did Sam Shepard's *Mad Dog Blues*, which is a really strange musical about people tripping on acid. It was my first experience with sound for the stage. I was told I was going to mix the show and the college rented some equipment from a man who worked in bars and did industrials. He had no theater experience and I had no sound experience. Together we were quite a team. I can't even imagine what it sounded like now, but I doubt it was very good. I do remember I really enjoyed it and it opened my eyes to a new world in theater. That year I also started the obligatory college band and bought a four-track recorder. Remember those? Four tracks on a cassette. I started getting more interested in sound, but I hadn't really figured out that there was a career in it.

After college I went to graduate school for Dramaturgy, which is basically a person who studies theater and helps explain the socio-economic environment that existed in the time of Ibsen's *A Doll's House* and stuff like that. I was into directing and writing and researching theater, but after my first year I realized it just wasn't what I wanted to do. The whole time I was in school I was working as a stagehand to earn money, and I realized that I enjoyed being a stagehand more than anything. So I dropped out of school and started a dance/theater company with some friends and played with my band in bars all over Texas and Arkansas and Oklahoma.

Those were the lost years and, man, were they fun. I also worked as a draftsman at an engineering firm, which had been another source of income for me since I was 13. At the ripe old age of 13, I had been introduced to AutoCad 1.0. Very few people knew how to use it at the time, and I figured it out and worked as a draftsman during the summers. After I dropped out of grad school, I fell back on drafting as a way to make money while I figured out what I was doing.

By this time, it was the early 1990s and sound had really taken off in theater in my area. My good friend from college, Patrick Pummill, was mixing musicals at a local theater and was soon to start touring, which would lead him to Broadway. I started following his path as it became clear that there was a career to be had. I found myself working at an avant-garde theater called the Undermain and became the sound mixer. My then girlfriend and now wife, Mollie, got me my first job in sound, and she loves to point out the fact that I owe my career to her, so props to Mollie. I loved mixing and I loved that little theater. I mixed punk rock musicals and crazy plays. The more I worked in sound, the more I understood how it brought everything together for me. Unlike any other technical discipline in theater, sound is organically involved in creating the performance. As a mixer you are part actor, part musician, part stagehand, part director, and part dramaturge, and it was a perfect fit for me.

Then I moved on to mix at the Dallas Theater Center and I met Curtis Craig, who really opened my eyes and changed my life. The Dallas Theater Center is a very nice regional theater with a 500-seat theater and an 800-seat warehouse space. We did some great work there and it was full of creativity. Curtis was a fantastic sound designer, and I was so impressed with the work he did on plays. He built great layered soundscapes; at the time, we were using minidisc players and some samplers. Can you believe that? Curtis is still a very close friend. He currently teaches at Penn State and anyone who gets to study with him is incredibly lucky. I have hired several of his graduates and they have all been of the highest caliber.

While at the Dallas Theater Center I stumbled upon a job posting that changed my life even more. I applied for a position as a touring sound person on a show called *Tap Dogs*.

I was so young and dumb and I had no clue how unqualified I was for the job. Fortunately for me, though, when I applied the tour just happened to be in Dallas. The designer was touring on the show; he needed to take a few months off to do another show and was having no luck finding anyone, so I met the designer for an interview and he hired me. As it happens so often in this business, I was offered a job and had to be on a plane a couple of days later and would be gone months. So I said my good-byes and became a touring mixer.

The designer's name was Daryl Lewis. He was from Australia, which meant he talked funny. Once I arrived at the theater in Springfield, Missouri, I think it took Daryl less than five minutes to size me up and realize what a mistake he had made in hiring me. I was completely in over my head. I went from mixing shows for 500 people to mixing shows for 3,000 people. I had never heard of Camlocks or G-Blocks and the Mackie I had become so proficient with at the Dallas Theater Center didn't have VCAs. I can only imagine what Daryl thought when I asked, "VCAs? What are those?" I arrived for the last day in Springfield and I watched the show that night and was blown away. I couldn't believe I was going to mix that show. Then we did a load-out, which was my first load-out; I still have a scar from it. When we were dropping the Yamaha PM4K, Daryl asked for hands all around the desk. I jumped in, and as we set it on the ground, Daryl warned us not to get our hands caught under the 600-pound desk. We placed it on the ground and as everyone else stood up, to my horror, I realized that my middle finger was stuck under the desk and there was a piece of sharp metal digging into my flesh. I cried out for a little help and my bloodied finger was released from the console's death grip. I learned an important lesson right then, which is you never drop a desk like that. Someone could get hurt. And now anytime I start to get a little uppity, I just look at the scar on my finger and that deflates my ego a little.

The next stop was St. Louis at the Fox, which is a 3,000-seat work of art. The first time I touched the desk was in front of 3,000 people and I was a nervous wreck. I never knew how much sweat could pour out of your palms. Daryl made me bring napkins out front every night to wipe the desk off after I mixed portions of the show. He was not happy with me. Basically, after each show he would say, "Well, that was pretty bad. I don't think this is going to work. Keep your bags packed because I will probably fire you tomorrow." And then I would go back to my room feeling like a complete failure. It was kind of like the scene in *The Princess Bride* where the Dread Pirate Roberts would tell Westley, "I'll most likely kill you in the morning." But luckily Daryl had a plane ticket back to Australia for another job and I was his only hope, so he kept working with me. I learned so much from Daryl; he was a great mixer and a great guy. By the time he left, I was doing a good job on the show, and I took care of the tour while he was gone with very few problems. After that I worked for Daryl for a couple of years.

With that change in my life, I began my life as a theater hobo of sorts. I didn't have a place to live for about seven years other than a bus or a hotel room. I lived on the road and bounced from one tour to another or to a summer stock or regional theater when I needed to. I did my time working for most of the touring production companies, and I toured mostly bus and truck one-nighter tours of musicals. Slowly I started moving up the ladder as a mixer and worked on bigger tours with longer sit-downs. Then I got a job on a union tour and received my union card. Next stop, New York City. When I arrived in New York, it was like I started over. I was back to mixing in basement theaters and at the Fringe Festival and making no money.

It took some time to make connections, but luckily my friend Patrick Pummill was mixing on Broadway by then, and he introduced me to Kai Harada and Tony Meola. I can't say enough about these two people. Plain and simple, Tony is one of my idols. To me, he is a master at his craft, and I fall right in line with his design aesthetic. Kai is one of the most talented and intelligent people I have ever met. After a couple of years doing this or that, I eventually started mixing as a sub on Broadway musicals. My first show was a Tony and Kai show, which was *A Christmas Carol* at Madison Square Garden. It is just a little 5,000-seat house and we did 15 shows a week for eight weeks. Then I moved on to other shows. At one point I was the sub mixer on four Broadway musicals. It was crazy, but I loved it. Mixing musicals is definitely my passion.

The next step for me was to become a Local One member, which took me almost seven years. For the past few years I have worked as a deck sound person, sub mixer, and monitor mixer as a house sound person. I also have moved into designing. I rarely design anything but musicals, but I have designed a fair number of musicals. To me, musicals and plays are completely different. Different aesthetics, different techniques, and different skill sets. I *can* do plays as a mixer, but I prefer musicals. I *can* design a play, but I prefer musicals.

So that's me in a nutshell. Now, how did I come to write this book? Well, it's a funny story, actually. Over the years I started developing software on the side, and I became a partner with Stage Research. Stage Research revolutionized playback for theater when they created SFX, which is playback and show control software. Before Stage Research we used minidiscs, CD players, and samplers for sound effects. Once SFX came along, we started using computers for our sound effects. It has made for vastly more complicated sound designs that are infinitely easier to run. I developed a program called ShowBuilder, which is a software solution for paperwork management in theater and is especially geared toward sound. Several years ago Carlton, one of the owners of Stage Research, asked me to create a hand-drafting sound template. He put me in touch with Fred Allen and Steve Shelley. These are the guys who created Field Templates, which is the company that makes hand-drafting templates for lighting. So, we worked together and put out a sound template that could be used to teach basic sound theory as well as be used for hand-drafting. Of course, no one hand-drafts anymore, but the template can be useful in production meetings if you want to do a quick sketch of your plan. Last year I was at USITT and ran into Steve in the hotel bar. Steve had just released the second edition of his book *A Practical Guide to Stage Lighting*. I was primed with a few pints, so we started chatting it up and I asked him about his book. The next thing I know, a drunken slur of a rant starts pouring out of my mouth about how there really hasn't been a book written about mixing. About being a theatrical mixer. There are books on design and electronics and the like, but no books on the process and career of being a mixer. Steve told me

I should write that book, and of course I blurted out, "That's a great idea," as I finished off another pint of Guinness.

Steve told me he would introduce me to his publisher, which he did, and the next thing I know I am telling Mollie I have to write a book. Of course, she thinks I am crazy, but she knows me and she knows there is no stopping me when I get some thought in my head. So now I find myself spending all my time thinking back over my career in sound and all of the lessons I have learned and all of the amazing people I have worked with. I want to share the knowledge that I have been fortunate enough to soak up. To me, being a sound mixer is a trade that has been passed down from one mixer to another. The history of sound in the theater is still relatively young. It is still possible to meet and talk with the pioneers of the field. How amazing is that? If you have never heard Abe Jacob give a talk, then you have truly missed out and need to find a way to hear him. I also consider myself a historian of theatrical sound. I am fascinated with old war stories and thoroughly enjoy knowing the lineage of mixers and designers. At some point in this book, maybe I will put together the genealogy of theatrical mixing. I feel like there is such a rich and interesting history, and I want to help preserve it so others can learn from it.

As sound has developed into an important and crucial part of theater, colleges have started teaching sound. I believe there is a place for these programs to offer a book that gives insight into this amazing field. I believe this book is for several types of people. Most obviously, it is for college students, but I also believe it is for people doing sound in little black box theaters or regional theaters. I am writing this for anyone who wants insight into the world of theatrical mixing at the highest level. I am focusing on musicals rather than plays because I believe they are very different skill sets, and it is vastly more difficult to mix a large-scale musical than even a large-scale play. For the most part, if you can mix a musical, you can move over and mix a play without an issue. The same can't be said of the reverse.

I started at the lowest level in theater and worked my way up. As I moved up the ladder, I learned what was possible at the next level. Once I reached the level of mixing a Broadway musical, which I consider the gold standard of theatrical mixing, I learned what is possible when you have a huge budget and plenty of tech time and previews to perfect your mix in a way you just can't do outside of Broadway. Now when I go back and work at a regional theater or black box 50-seat theater, I take the lessons I have learned with me. I don't lower my standards or expectations or my techniques. I work within the limitations of the theater, but I have a bag of tricks that I use to push that theater as much as I can. I bring knowledge that I know I didn't possess when I was at that level, and I am writing this book so anyone at any level can hopefully take bits and pieces of knowledge used to mix on Broadway and apply it to his or her situation. I am so passionate about theatrical sound and mixing musicals. I think we, as mixers, are such a special part of a show and I want theatrical sound to continue to evolve as the most artistic technical aspect of technical theater. I want to do anything I can to help people become better mixers. I am writing this book because I wish it had existed when I finished my first load-out on *Tap Dogs*. So now I find myself sitting in coffee shops tapping away about sound. I guess it is a good thing I studied English in college so I know, where to put, commas. ☺

How is this book laid out? What I want to do is lay out the book in the logical progression of a typical Broadway musical. I have different parts for the book that represent the different major parts of production on a musical. The first part is the general overview of what the career path is and the different players involved. After that, the parts move through the beginning to the end of a production. I break the parts into chapters that highlight the major areas of that part of the production. Some chapters have sub-sections to further break down the concepts of that part of the production. I am laying it out this way so it is easy for you to jump to an area where you need reference to a certain issue. The point of the structure of the book is to make you feel like you have gone from the beginning to the end of the process of a Broadway musical. After that, hopefully you will have added to your bag of tricks and will be able to apply some of these lessons to your process.

I hope you will see this book as a companion for books like Bob McCarthy's *Sound Systems: Design and Optimization:*

Modern Techniques and Tools for Sound System Design and Alignment, Second Edition, the classic Yamaha *Sound Reinforcement Handbook*, or John Leonard's book *Theatre Sound*. This book is intended to complement these and fill a hole I think exists about mixing. I don't plan to go very deep into design theory or system design in this book. People smarter than I have done a better job with that than I could ever hope to. Instead, I will keep my thoughts mostly to having your hands on the faders. Remember . . . up is louder.

As for the last question: if a speaker feeds back in the woods, will anyone hear it? The answer is yes. Somewhere, some sound person will feel a disturbance in the force and grab his ears and not know why. Ears are muscles, and the more you use them, the stronger they become. This has a positive and negative effect. You will be better at your job, but you will be miserable when you hear a faint 60-cycle coming through in the speakers at your local coffee shop. Everyone else will be laughing and having a great time, and you will be in agonizing pain. But it is worth it to have the opportunity to mix someone like Sutton Foster singing a love song. It is completely worth it.

Now let's get started.

THE CAREER PATH

WHAT DOES IT MEAN TO BE A THEATRICAL MIXER?

It is late in the second act. The story has been well established and the characters are now very familiar to the audience. We are way past exposition and just past the climax. The helicopter has already flown over the audience and the cast has stormed the Bastille and the guy in the half mask has already driven his fake boat through the sewers. And it has all been very magical. The lighting and the smoke and the projection have fleshed out the scene and the orchestra has played at full tilt until the room is ready to blow. Then the stage goes to a blackout. The fireworks and spectacle of the show are over and the audience applauds wildly at the end. The lights slowly creep up and it takes a while for the audience's eyes to adjust. A small pool of light appears upstage left and in that pool of light is one actress. It is the female lead and it is her big solo, in which she accepts her fate or decides to turn it around and change her fate. Just about every show has a moment like this, and this show is no exception.

The song starts off as a whisper. The orchestration is thin, no more than two violins and a cello. Then it starts to grow. More instruments are added. She stands up. She hits a big note as the orchestra swells and then it all immediately shrinks back down to a couple of strings. Then it starts to build again, but this time there are more instruments, and now she is walking downstage. Lights are coming up all over the stage and the scenery is quickly removed, leaving nothing more than this one actress belting out her song with the orchestra of 24 musicians supporting her. She hits a crescendo note, then a second of silence, and then she explodes into a huge crescendo. The orchestra follows her and the song ends to thunderous applause. It is massive and ear-crushing applause. This is the release of all the emotion in the show. It is bigger than any spectacle in the show, but it also depends on that spectacle to get the audience to the right emotional place. And their reaction is huge.

But there is a problem. At the back of the house, in the mix position, the sound board operator for the show sits in absolute silence with his hands on the faders. He is mentally beating himself up over the last song. Sure, the crowd reaction was huge, but he knows it was off. He knows that on a night when it is perfect, he can hear people quietly start to cry after the first crescendo. He knows that when it all goes right, there is a standing ovation at the end of the number. He knows that the applause normally goes on long enough for him to stretch for a second and drink some water before the next scene. But tonight, even though the audience went nuts at the end, he knows he didn't hear anyone get choked up, the audience didn't stand at the end, and he didn't get a sip of water. And he is not happy about it.

As he mixes the next scene, he goes back over it in his head. The audience is completely satisfied and they have no idea that anything was wrong, but the mixer has mixed the show over 200 times and he knows. He wants to know what he did or did not do that changed the response. Did he push the big spectacle scene too loud? If he did, that could've left patrons with fatigued ears not ready for the song. Did he start her too quietly? If he did, the audience might have had trouble hearing her at the top, and that may have thrown them out of the moment. Was she too loud at the top? If so, it wouldn't have been enough of a change to set the audience up for the rest of the song. After going over it again and again, he realizes what went wrong. After the first big crescendo he didn't pull everything down enough, so that left him with nowhere to go for the big finale of the song. The audience has no clue, but *he* knows their reaction could have been bigger.

That is what it is like to be a live theatrical mixer. It has almost nothing to do with knowing the model numbers of every speaker and microphone. It has almost nothing to do with speaker placement and system equalization (EQ). It is about the symbiotic relationship between the amplified sound of the show and the audience reaction. It is about understanding the arc of a show and the arc of a song. It is about manipulation. There are mixers who cannot program a Lexicon 480L reverb unit but can mix with such emotion and ease that you forget the mics and speakers even exist. There are also amazing sound people who can program that Lexicon and field strip it and rebuild it, and yet have no interest in mixing.

Mixing is an art like no other technical aspect of technical theater. It cannot be simplified to a push of the button. Mixing is dependent on several shifting factors. An actor is not feeling well, so she sings differently. A substitute, or sub, musician is in the pit and he plays louder. The audience is smaller than normal. The weather has changed and the room sounds different. It has been said that mixing is like playing a piano in which the notes are not linearly arranged, you have no clue where middle C is, and you have to walk up to it and play it perfectly. Francis Elers, who has mixed on Broadway for the last 15 years and has mixed shows including *Rent*, says mixing is like freestyle rock climbing with no safety in place. One wrong move and you are going to fall hard. Jordan Pankin, another long-time Broadway mixer who is currently mixing *Wicked*, explained mixing as a boxing match. He said you walk up to the desk, stand toe to toe with it for three hours, and see who wins.

The best mixers out there are the ones who embrace the idea that the job of the mixer is to become part of the story and to manipulate the audience as much as possible. You still need to understand a sound system and all of the basic physics of sound, but your real assets as a mixer are your ears and your ability to move faders. Some call it being a "fader jockey," and that term is not disrespectful. If you develop these skills you will be very valuable, especially to designers. Designers have to be very selective when choosing their mixers. When a designer is doing a musical, he knows that his design, in the end, will only sound as good as the person moving the faders around can make it sound. There will always be a very human element involved with the sound of a musical.

Actors are just as aware of the importance of having a highly skilled mixer. My first Broadway mixing position was for *Man of La Mancha* starring Brian Stokes Mitchell at the Martin Beck Theater, which is now the Al Hirschfeld Theater. Mr. Mitchell is an absolutely incredible actor with an amazing voice. He won a Tony in 2000 for his role in *Kiss Me, Kate* and is also currently the chairman of the board of the Actors Fund. Mixing him singing "The Impossible Dream" is one of the highlights of my career. I was being trained to be a sub mixer by Jordan Pankin, who was the full-time mixer. A sub mixer is the emergency cover mixer and the sub usually mixes one to two shows a week. Every musical on Broadway has at least one sub, and some shows have two or three subs. On my first day, Jordan took me to introduce me to Mr. Mitchell. Jordan told me that Mr. Mitchell had final approval over the mixer. Basically, if he did not like the way I mixed the show, I would be out of a job. Mr. Mitchell was very nice and, luckily, I kept my job.

It is an important lesson to learn. Your job as a mixer is not relegated to the sound in the house. I have worked on shows in which the sound in the house was absolutely fantastic, but the sound onstage was not working for the actors. On one show in particular there was nothing but compliments in the house and nothing but complaints backstage. No matter what we tried, we just could not make the actors happy. It is a tough balance to find settings for the foldback that make the actors happy and don't sacrifice the sound in the house. (*Foldback* is another word for the monitor mix onstage. The word comes from the fact that you are folding part of the sound back to the stage.) Finally, after months and months of working, the monitor issues were resolved. It is a horrible feeling because you want to help the actors, but at a certain point physics works against you, and there is apparently no good way to explain that to actors and have them accept that this is the best it can possibly get.

I designed and mixed a show once that required me to work with an extremely challenging actor regarding the monitor mix. When we started the tech process, the levels onstage were normal levels for a musical, but the cast was not happy. As we moved into previews I was asked to turn it up and up and up. I tried to keep it from getting out of control, but it was a losing battle. Finally, I was asked by the producers to give the actor whatever he wanted. I argued that the levels he wanted onstage would greatly change the sound in the house, which everyone was pleased with, but the decision was made that it was more important to make the cast happy. So I cranked it. The taps onstage peaked at 110dB, which is louder than a chainsaw. The crew complained and started wearing earplugs. They even posted OSHA signs warning of the hearing damage that can be caused by extremely loud volume. The sound onstage was so loud that there were times when the front-of-house system could be turned off and it would still be too loud. And yet it still wasn't loud enough for the cast.

After having a long conversation with the actor about the monitors onstage and what could and could not be done, he looked at me and said, "Why is it that it sounds so perfect in my head?" I honestly did not know what to say to him, but he very succinctly summed up the actor/foldback dilemma. The actors usually want it to sound onstage like it does in the house. Actually, they want it to sound like it would in a movie. They look at you skeptically when you explain that if you put an omnidirectional microphone into the speakers they are standing near, then the mic will probably feedback before they hear themselves. It is even harder to explain to them that, even if it doesn't feedback, it will affect the sound in the house because of two very important reasons. One is that their mic will be picking up their voice along with their voice from the foldback speakers with a huge delay, and that will cause them to sound hollow and muddy. The second reason is that the speakers onstage will bleed into the house and muddy up the sound in the house. And don't even try to explain to them that they will hear themselves with a delay that will actually cancel out their voice and could possibly make them sound even quieter. None of that matters. They know in their heads what they need it to sound like onstage and that is what they expect.

But there is hope. Luckily, the main concern of most actors is how they sound in the house. They want to sound amazing to the audience. However, something happens with almost every musical that puts the cast on edge, when the cast takes the stage for the sitzprobe or wandelprobe. (A *sitzprobe* is when the cast sits onstage in chairs and runs through the music in the show for the first time with the entire orchestra. It is also basically the sound check for the sound department. A *wandelprobe* is the same, except the cast is allowed to walk the stage and do blocking and choreography.) Since this is the first time, there is a ton of work for the sound department to do. Inevitably the balance is not right for the cast, and it takes time to get it right. All the cast knows is that it does not sound good onstage. They can't hear themselves or the rest of the cast, and the orchestra is not balanced at all. Naturally, their first fear is that they are going to sound bad in the house, and their second fear is that they are not going to get what they need.

Actors are putting a great deal of trust in the sound mixer. They are basically putting their talent in your hands, so it is important to gain their trust and to let them know that you are doing everything you can to make them sound incredible for the audience. You have to allow them the time to understand that you are working as hard as you can to make them sound as good as possible. Once they find out that it does sound great in the house, they will usually relax about the sound onstage. They will still need help, but they will be more patient. If they find out it is a train wreck in the house as well as onstage, then it will be a long and miserable process. In the end, your saving grace is how good it sounds in the house.

This does not mean you don't do everything you can to make it sound good onstage. I have worked shows where there are almost as many speakers upstage of the proscenium line as down. A happy cast makes for a much better experience. Sometimes it is possible to give the cast everything they need, and sometimes it isn't. Sometimes you have to find a compromise between what they need and what is going to affect the sound in the house. But it takes time and planning and patience to achieve this. Sometimes having an actor feel comfortable so that a good performance is achieved is better than a perfect mix in the house and a lousy performance. Sometimes the compromise is the best choice. Sometimes you have no choice but to sacrifice the sound in the house for the sound onstage to best serve the show. However, getting to the final sound onstage is like running a marathon. You have to be ready to run uphill with people pouring oil on the track.

I designed a short run of *Cinderella* and my mixer was Chad Parsley, who has mixed *Jersey Boys*, *Avenue Q*, and *Spamalot*. We were in the middle of the wandelprobe and I walked up onstage, stood with the actors during a couple of songs, and worked with Chad to set the levels onstage. At the end of a song, I talked to the actors and told them to let me know what they needed. I of course received the usual answer: "I need more of me. More of me. More of me." I explained that I would do the best that I could and to give us some time to get everything dialed in. As I left the stage and walked toward the back of the house, the director stopped me and asked if everything was okay. I said everything was fine and explained that the cast was trying to be patient with me, but they would hate me in another hour for not turning their mics on in the monitors, and then slowly but surely they would go out and listen in the house and hear how good it sounded and would then calm down. She smiled and said it sounded normal, and told me to let her know if I needed any help.

As you can see, your job as a mixer has several layers. The first layer is a thick skin. You have to be aware of the incredible vulnerability of the actors you are working with. There may be times when everyone in the room is annoyed with you, and you have to acknowledge that they have every right to be. You have to hold it together and fix whatever is annoying them. You have to exude a confidence that everything is going according to plan, even if the console is on fire. The next layer is the artistic layer. You are there to amplify the words and the music. You facilitate the communication of story between the actor and audience as well. You are there to help carve out the emotional content of the show. You are also there to make the actors as comfortable as possible. But that's not all. The next layer is the business layer. You are also there to train a sub how to mix the show, and you are responsible for ordering the perishables, such as batteries and tape. You are in charge of working with the rental house to replace broken gear, and you come to work no matter how sick you are unless you have a sub to cover you.

There is a certain amount of confidence required to be a successful mixer, but it has to be couched with a decent balance of humility. You have to be confident enough to make quick decisions and implement them with ease. You have to know what knob to grab and how to EQ a mic. You have to have the confidence to know that your EQ is correct, but you also must have the humility to know that you are there to serve the designer and his design. As a mixer, you do not have to approve of every decision a designer makes; after all, sound is very subjective. You may like more hi-hat than the designer, and that is fine. As a mixer, you have to respect the decision of the designer and mix it the way you are told to mix it. I don't think I have ever mixed a show in which I agreed with every sound coming out of the speakers or every choice made by the designer, but that doesn't mean I disapproved of the overall sound of the shows. So what if you don't like a certain reverb? The director and producer and designer like it. That is all you need to know. It is a huge challenge as a mixer, and it is why designers gravitate toward mixers who have very similar ears to their own.

There is also a certain Zen place that an experienced mixer reaches while mixing a show. When you are new to the world of mixing musicals, it can be completely overwhelming. I remember my first big show. The designer/mixer made fun of me for sweating all over the VCA section on the Yamaha PM4000. What can I say? I was nervous. I was learning to mix a show, a huge show, live and in front of an audience of 3,000 people. I wasn't doing a very good job, either, and I was very close to being fired. Luckily I pulled it off and toured on the show for several years. When you are learning a mix, it seems so fast. There does not seem to be any time even to breathe, let alone fix a problem on the fly, but once you learn the mix and you've mixed the show 50 or 60 times, it all slows to a crawl. That one scene that seemed impossibly fast is now creeping along and, not only can you mix it, you can also fix a problem and write down notes and call backstage to find out what's for dinner. Once you do several years and several dozen shows, mixing in general just becomes easier. It becomes second nature. You get to the point where you can just walk up and throw faders and mix a musical cold with no run-through and without any panic or nerves getting in the way. It's just you standing toe to toe with the faders: all the layers disappear and you get lost in the story and you mix.

HISTORY OF THEATRICAL SOUND

There is an old Latin saying, "nanos gigantium humeris ins-identes," which translates to "dwarfs standing on the shoulders of giants." Of course, this has become a Western metaphor meaning someone who develops future intellectual pursuits by understanding the research and works created by notable thinkers of the past. In the world of theatrical sound, and especially theatrical mixing, this metaphor could not be more accurate. We are all dwarfs climbing up inch by inch to reach the shoulders of those who came before us. Hopefully, when we get there we add something to the field so it continues to evolve and grow. The difference between the mythological origins of this statement and theatrical sound is that some of our giants are still walking among us. Our field is so young that we have the advantage of being able to see the creation and growth and, hopefully, some of the future of this trade. In order to attain the majestic views from atop the shoulders of these pioneers, we need to look back at where we came from.

One brisk morning I found myself loading-in to a theater in Macon, Georgia, for a one-nighter. It was a beautiful old theater and the crew was very proud to talk about its history. (The theater even sold a coffee table book about the history of southern vaudeville houses, which I purchased and slowly checked the theaters off as I visited them.) The crew enjoyed showing off their trapdoors, which were cut specifically for Houdini. As a touring stagehand, I was fascinated by the thought of people touring in the 1920s, and that someone had to advance the theater and tell them where to cut the trapdoors.

In the afternoon, once the show was loaded-in, the house carpenter told us there was a haunted tour of the theater and asked if we would like to take the tour. Who could resist? So the house crew scattered, and we were led through the theater by the house carpenter. Every once in a while, a house stagehand would pop out of a wall through a secret door, or a hand would creep out of a secret hole and grab a shoulder. It was great fun. The final stop on the tour was the thunderclap room.

We entered a large eerie space (Lights 235 and Sound 110—Go). It was directly above the seating chamber of the theater. The house carpenter pointed out a red splotchy outline of a body on the wall. Apparently, decades earlier someone who worked for the theater had committed suicide in this room and his dead, rotting body baked in Georgia's sweltering summer heat until it swelled up and exploded, leaving a permanent stain on the wall. They had to remove a wall and use a crane to remove the remains. It was horribly sad and extremely gruesome, and a dramatic ending to the haunted tour. It was explained to us that this was the thunderclap room where stagehands used to create sound effects for the plays. Of course, the most common sound was banging a piece of sheet metal to create a thunderclap, which was amplified by the space of the room and forced out of the tiny opening that pointed down at the audience. It was amazing. We were standing in an acoustic speaker, and it showed me how far sound had come in a short time.

Theatrical sound was all foley in the beginning. *Foley* is the art of making sound effects live with objects. It is a very honest way of adding sound to a play and it is still used today on stage and screen. There are times when someone standing offstage slamming a door sounds infinitely better than any recorded sound of a door slam coming from a speaker offstage. Live gunshots are always better. Doorbells and phone rings . . . definitely better. But there are limits. If you need an airplane to fly by, it is a little impractical to actually fly a plane through the theater, so it was inevitable that recorded sounds would become essential to theatrical sound. The first production credited with using a recorded sound cue, as cited by Bertolt Brecht in Belgium, was a play by Tolstoy about Rasputin in 1927, directed by Erwin Piscator. Piscator was a Dadaist and really pushed the envelope of the technical elements in theater. In this case, he needed a sound effect that included a recording of Vladimir Lenin's voice. From this humble Dadaist, Socialist beginning, recorded sound became a part of theater, although it wouldn't take hold in mainstream commercial Broadway theater for another couple of decades.

Recorded sound cues slowly crept onto the stage, but it wasn't until the 1950s, when movie directors began directing stage plays, that recorded sound cues became more the norm. There were some who took notice of the imminent rise of sound in theater and established themselves at the forefront of the movement. Masque Sound is a rental shop that specializes in Broadway theater. Three Broadway stagehands started Masque Sound in 1936 when they purchased a small record label and began to produce sound effects records and rent sound equipment to Broadway shows. In the early 1950s, Masque Sound was the first company to adopt tape technology for the theater. Since then, Masque Sound has expanded along with the growth of the theatrical sound industry. Sound Associates is another sound rental shop that was there at the beginning. Sound Associates began in 1946 and has grown as well to be one of the major players in theatrical sound for touring and Broadway.

Even though shows were starting to use recorded sounds and people were creating the cues for the shows, at that time there was no sound designer position in theater. The first person known to have received a credit as sound designer on the poster and in the program alongside the lighting and scene designers was David Collison, at London's Lyric Theatre, Hammersmith, in 1959. Mr. Collison started in theater as a stage manager and went on to become a very successful mixer and designer. The first person credited as a sound designer on a Broadway show was Jack Mann for *Show Girl* in 1961. Jack Mann started out in theater as a performer. He was the Master of Ceremonies in *The Green Pastures* in 1935. He transitioned over to sound in the 1960s and designed classic musicals such as *Company, A Little Night Music, Follies, Sweeney Todd*, and *West Side Story*. It wasn't until 1968 that a regional theater gave Dan Dugan sound design credit at the American Conservatory Theater (A.C.T.) in San Francisco, and with that it seems sound finally had a legitimate place in theater.

Theatrical sound was something very different back then than what we are used to now. Abe Jacob, who is considered the godfather of theatrical sound, was interviewed in September 2000 by David Johnson in *Live Design* magazine:

> *Theatre sound design at the time was handled by the show electrician and stage manager. For the most part it was merely a matter of putting in the dressing-room page system and five mics across the front of the stage. The stage manager would tell the electrician, who was hired to put the board in, that this is the mark you put the knobs at, and when the director's in the house, you put them up or down a little bit—and that was sound. Now certainly there were individual performers prior to that time who wore wireless mics, so that was another thing that was turned on and off when the star was onstage, but again it was brought up to a mark and left.*
>
> (David Johnson, "Meet Abe Jacob," *Live Design Online*, September 2000, available at http://livedesignon line.com/mag/show_business_abe_jacob/)

Mary MacGregor, a Broadway sound person who has been working on Broadway for the past 30-plus years, remembers what it was like to work sound on Broadway in the early 1970s:

> *Sometimes all we would need to do was hang a shotgun mic on the first electric. That mic would feed the dressing room system and the front of house. I remember having to ask for a place on the pipe and a separate pick point for my cable so it wouldn't buzz and the head electrician was so annoyed that he couldn't bundle my cable in with his. We were only hanging one mic and we were already in the way. They didn't know what was coming next.*

Abe Jacob marks a very distinct change in theatrical sound. He got his start in San Francisco mixing sound for such 1960s rock stars as Jimi Hendrix, the Mamas and the Papas, and Peter, Paul, and Mary. He also designed the sound system for the Monterey Pop Festival, among other things. He became involved with theatrical sound in the early 1970s when he worked on a production of *Hair* outside of New York. As we all know, *Hair* is more of a modern rock musical than a show like *Show Boat*. It is a completely different sensibility and one that requires more skill and dedication to the mix than an electrician sitting backstage setting recorded levels can give a show.

Mr. Jacob explains how he started in theater in the 2000 *Live Design* interview:

> It was easy, because no one else was doing it. Before theatre, I was doing a lot of concert sound. The whole idea was that I was doing concert sound and then started doing theatrical work on the West Coast, and through that got involved with other productions of Hair *that happened outside of New York. And I ended up in New York at the time of the previews of* Jesus Christ Superstar, *when they had some sound problems as well as other technical problems. I happened to be at the theatre at one of the cancelled performances and re-ran into Tom O'Horgan, whom I'd worked with on one of the* Hair *companies outside New York, and he said, could you help? I was in town for a few days and did, and I guess I've been here ever since. That was early 1970. And word just got around. We did* Superstar, *and it worked, and from that point on, various management offices would get in touch with me. I was actually planning to move to New York because I was going to manage Electric Lady Studios, Jimi Hendrix's recording studio down in the Village. That was the secure job that I had to come here to so I could live, and then theatre was on top of that. But that lasted about six months and the theatre became much more lucrative.*

At that time there was no position for a sound operator, but that was about to change. Mr. Jacob explains:

> I brought in some people who had mixed concerts with me to run the shows now that we had the opportunity to design. And the things I insisted on were that the operator, since he's running the show, needs to be in the audience where he can hear what the audience is hearing and away from backstage next to the dimmer boards. That meant another person to

do it, and so that's how the sound operator got to be a part of the crew. And it established sound design as a credit on the title page. I suppose the first big musical we did from scratch, where I was hired before they went into rehearsal, was Pippin (directed) by Bob Fosse. And then it just seemed to go on from there. I was fortunate in that I was able to work with directors who shaped the American musical theatre in the early 70s through the 80s—Michael Bennett on Seesaw and A Chorus Line, Bob Fosse on Pippin, Chicago, and Big Deal, and Gower Champion on Mack and Mabel and Rockabye Hamlet.

Even though directors and producers had warmed up to the idea of sound as an important part of theater and to the need for a designer and an operator, the critics were not so easily swayed. According to Mr. Jacob, "it was almost immediate that critics started making comments about the sound in the theatre, and that it was going to bring about the death of the American musical as we knew it." Mr. Jacob believes that one reason sound started to become accepted by audiences, though, was because of the invention of the Sony Walkman®, which changed the way individuals listened to music and what they expected to hear in a live show. With the invention of the Walkman came the ability for people to walk around listening to music through lightweight headphones, which heightened the listening experience and increased their awareness of the quality of the sound they were listening to. Not only that, but movie sound began to improve and raised the bar for what people expected to hear. As their awareness of sound quality increased, so did their expectations for live performances. Now it was up to the theater to live up to these expectations.

Abe Jacob is considered so important to theatrical sound for several reasons. He brought studio and rock concert techniques to the theater. He helped solidify the sound designer position as well as the sound operator position, and helped carve out a mix position for us to work. His resume is incredibly impressive. He designed *Hair, Jesus Christ Superstar, Beatlemania*, the original *The Rocky Horror Show, A Chorus Line*, and *Cats*. Mr. Jacob also hired a string of mixers who themselves became the next wave of Broadway sound designers, including Otts Munderloh, Tony Meola, Steve Canyon Kennedy, Jon Weston, Lew Mead, Duncan Edwards, and Kurt Fisher. Abe Jacob is to theatrical sound what Paul Erdös is to mathematics. Erdös is considered to be so influential to math that mathematicians have what they call an Erdös number, which indicates how many degrees from Erdös you are. If you published a paper with him you are a 1, if you published with someone who published with him you are a 2, and so on. Theatrical sound should have a Jacob number for mixers. (Since I have mixed for several of Mr. Jacob's mixers, that would make me a second generation, or a 2.)

DIFFERENT POSITIONS IN THEATRICAL SOUND

In any profession, it is important to understand the hierarchy you are dealing with. It is also important to understand the roles available and the duties and expectations of these roles. Without a strong understanding of this, it is harder to know what personal goals to set and how to achieve those goals. It is also important to understand the positions, and the expectations of each position, so that lines can be established that are not crossed. Theatrical sound definitely has a hierarchy and roles that are well defined. There are times when several roles can merge into one, but there are often times when the positions are isolated and defined. There are times when there is only one position, which really simplifies the hierarchy, but the reality on Broadway is a pretty structured set of roles with an established chain of command. Knowing and understanding this structure can be crucial to success in the business.

Designer

The role of the sound designer seems obvious enough, although when I tell someone I am the sound designer on a show I usually get a blank stare and a smiling nod. It is easy for people, even theater people, to understand the other design fields in theater, but sound is just not that tangible. With the lights or the sets or the costumes, people can see the design. They can touch it. Sound lives in the air, and it just isn't obvious to people that it takes a plan and a design to know how to move that air around a theater. I designed a musical at a regional theater once that had no sound cues.

The show was a straight up musical with two women and a country band onstage. I was really proud of the sound of the show, and I even received a good mention in the review, yet as the lighting designer was saying good-bye to me he made the comment, "It was great working with you. Maybe next time I will get to see you actually design." It stopped me dead in my tracks. To many individuals in the industry, it would appear that I had done nothing because there were no sound cues. This understanding of sound design is more common than not. So what is the job of the designer?

To understand the job of the sound designer, we have to look at the different permutations of the role. The first thing to understand is that there are two major categories for sound design, and each is ultimately responsible for the aural environment of the show. The first is a *musical theater sound designer* and the second is a *straight play sound designer*. There are massively different techniques, skills, and sensibilities required for these two distinct styles of sound design. It is similar to the difference between a pastry chef and a savory chef. Both are chefs, both can cook, and both can cross over to the other style, but both have a strength and a preference for one style over the other. A straight play sound designer is very often a musician or a composer. This type of designer specializes in building soundscapes and works within the script to find ways to heighten the emotion of the show using sound and/or music. Some great examples of this type of sound designer are John Gromada, Dan Moses Schreier, and John Leonard.

A musical theater sound designer does not have to be a musician or a composer. In fact, many musical theater sound designers are mixers or former mixers. This type of designer specializes in manipulating the sounds made by others to enhance the emotions of the show without being noticed. A musical theater sound designer usually tries hard to remain invisible, because if the audience is being drawn to the speakers, then they are being pulled out of the show and the suspension of disbelief is gone. Some great examples of this type of sound designer are Tony Meola, Brian Ronan, and Steve Kennedy.

This is definitely not to say that there can't be a designer who does both. Dan Moses Schreier is a perfect example of a designer who crosses over. He is a wonderful composer and an amazing designer of straight plays, and a very skilled musical theater designer. But it is important to understand that these are very different styles of sound design. Designing a musical usually requires knowledge of wireless mics and band mic'ing techniques, as well as an understanding that you are there to help bring the composer's sound to reality. Designing a musical requires an understanding that you are a vessel for someone else's aural creation more than a vessel for your own. Of course, that does not mean there is no creativity involved in designing a musical, but rather that it is more collaborative, with more cooks in the kitchen when it is a musical.

Inside of each of these styles of design, we can further break down the role of a sound designer to what has traditionally been called *technical sound design* and *conceptual sound design*. These are equally important and equally challenging aspects of sound design. This distinction started in the 1960s and defines the duality inherent in sound design. Often both roles are accomplished by one person, but there are times when the roles are covered by two people.

Technical sound design is also called *theater sound system design* by the United States Institute for Theatre Technology's (USITT) Sound Design Commission. Technical sound design deals with the nuts and bolts. This part of design is all about the gear and what will be needed to amplify the show so everyone can hear it. The technical aspect deals with picking the right amps and speakers and deciding where to hang the speakers. It also deals with choosing a console and processing that will accomplish what is needed. It also deals with EQ'ing the room. To EQ a room, a designer can use a variety of tools, such as Smaart or Systune or Meyer Sound SIM. Technical sound design also means planning out the cabling and racking of the equipment.

Conceptual sound design is also called *theater sound score design* by the USITT. Conceptual sound design is definitely the more creative aspect of sound design. In straight plays, this is the part of the design that selects or composes the music. In musical sound design, this is when the designer creates a vocal reverb for a specific moment. In both types of design, the designer is working on the conceptual design when he builds sound effects and works on reverbs and vocal effects. This is when the designer works with the director or composer to understand what their needs and goals are and finds ways to accent the emotions of the show. There are many straight play sound designers who are accomplished composers and not technical designers, and there are many musical designers who are accomplished technicians and not composers.

In the end, both aspects of sound design are crucial. If a show has a conceptual-only designer, then someone will have to fill the void on the technical side. This is much more common in straight plays than in musicals. A musical by its very nature is much more about the system than anything else, and there is already a composer for the show, so the conceptual needs for the show are different than for a straight play. The goal, however, is always the same, which is not just to make the show sound good, but to make the show happen—that is the real job of the sound designer. Tony Meola, the sound designer for *Wicked*, taught me a valuable lesson about what the role of the sound designer is. From him, I learned that sound design is not just about how your show sounds. You also have to be good at the politics. Sometimes I say sound design is 90% politics and 10% how it sounds. You can have the best-sounding show ever, but if you went way over your budget and had screaming arguments with the director and producer and worked the crew through every meal break, you probably are not going to be asked back. However, if you produced a good-sounding show under budget while being pleasant to everyone, then you will probably work for those people again.

Perhaps the most important aspect of being a sound designer is being the person who can pull it all together. The designer has been hired as the person with the expertise to make the show happen. In the case of a musical designer, that means the designer may have to explain why there are wireless mic problems or why the mixer is having trouble mixing a scene. The musical designer may need to explain why the vocals can't go in the monitors or why the hats are affecting the sound of the show. Being a designer is not about knowing the answer, but knowing how to get it and how to explain it so everyone understands. Brian Ronan, the Tony award-winning sound designer of *Book of Mormon* and *American Idiot*, is one of the best and most unassuming designers I have ever worked with, and he excels in solving problems. If he encounters a problem and doesn't know how to fix it, he will throw the net out and fish until he finds it. He is very open to opinions from his mixers, and the result is that he creates an accepting and collaborative environment—and his shows sound fantastic.

When designing musical theater, it is very helpful for the designer to have an understanding of the other positions and techniques involved. It is helpful for a musical theater designer to know how to mix, and historically most musical theater sound designers on Broadway have been mixers. Understanding how to mix a musical goes a long way to making a show sound the way a designer wants it to sound. It is also helpful to understand wireless mic techniques. It is good to know the process of prepping mics and hiding them. For all sound designers, it is crucial that they understand that part of their job is to create an environment where everyone below them can succeed. It is the designer's job to establish a good rapport with the other designers and to gather the information that is needed for the others to do their work. If a sound designer wants to be successful, it is his job to make everyone who works for him successful. A failure on the designer's team reflects directly on the designer.

Associate Designer

The associate designer is just under the designer in order of importance. The designer hires the associate, but not all shows have an associate designer. A person becomes an associate for a designer by working with the designer repeatedly and gaining the designer's trust. An associate should fully understand the designer's aesthetic and system preferences and be able to replicate what the designer likes. An associate is an acceptable stand-in for the designer and has been given permission by the designer to make decisions in the designer's absence. There are times when the designer cannot be at the full tech of a show, so his associate will take the lead and be the designer's representative whenever the designer is not around.

It is very challenging to become an associate. Usually the associate has mixed for the designer and has built and loaded shows in for the designer repeatedly. It is possible that the associate worked his way up by assisting the designer as well. Kai Harada is a great example of an associate sound designer. Kai is the associate designer for Tony Meola on *Wicked*. Kai worked for Tony for years as an assistant and did all of the paperwork an assistant does very well. Over time, Kai grew to the point where he could put together the paperwork for a show designed by Tony with very little input from Tony, and Tony grew to trust Kai's ear and his judgment. Then Kai moved from the assistant on shows to the associate on shows.

Being the associate on *Wicked* means that Kai has been given the authority to speak in place of the designer. He can fly out to a production and give design notes to the mixer. He can also make decisions on the direction the system needs to go on the tours or on Broadway. Since *Wicked* opened, there have been several productions of it around the world, and Kai has worked on all of them to keep the shows as consistent as possible. With so many productions of *Wicked* having opened, at times Tony is not available, in which case Kai, as the associate, acts and is treated as the voice of the designer. Now Kai has moved to designing his own shows, which is the logical progression of an associate.

Andrew Keister is another great example of an associate. Andrew is Steve Kennedy's associate. Andrew worked for years as a mixer for Steve. Andrew was a mixer and mixed *Hairspray* on Broadway, among other shows. He began assisting Steve as well as mixing for him, and soon he became Steve's associate. Andrew is the associate on *Jersey Boys*, another show that has been remounted all over the world.

Assistant Designer

The assistant designer is one step lower than the associate. A show may not always have an associate but almost always has an assistant. An assistant's job is usually not a high-paying one. It is intended for people learning the ropes or working their way up in the business. This is not always the case, however. There are assistants who have done several shows and are very good at what they do. Those assistants will usually be paid more because of their experience. Some assistants make a very good living bouncing from one designer to another. Those assistants are incredibly skilled at what they do, and they make it look easy.

The job of the assistant is very similar to that of the associate: to produce paperwork and answer questions in place of the designer. One difference between the associate and the assistant is that, whereas the associate has been granted license to make design decisions in place of the designer when needed, the assistant has not been given that authority. The assistant is more of a conduit to gather questions, bring them to the designer to get answers, and report back with those answers. Of course, this is not always the case, as the line between assistant and associate can blur based on experience and the history the assistant has with the designer. If an assistant has done half a dozen shows with the designer, he has probably reached the point where the designer trusts the assistant enough to take more of the reins.

When I design shows and I have an assistant, I consider our first show together as training. I invest time in explaining to the assistant how I like things done and why. I try to hand that person a system and paperwork that has been roughly thought out. Whenever I can, I surround the assistant with people who have worked with me before. The hope is that after the assistant has done a show for me, then the next time that person will be able to take on more of the duties of the assistant. After two or three shows, that person should know me and my quirks well enough to take the reins from the beginning and complete the project with little help from me.

Another way this can work is by new assistants working with old assistants or associates. There are times when a big-name designer is too busy and has his assistants and associates all working on projects, and so has to go outside the "family" to find an assistant for this other project. That is when you end up with someone with little Broadway experience or fresh out of college working as an assistant. When this happens, a lot of weight is put on the shoulders of this new assistant. Typically, the new assistant is given paperwork from older shows from one of the other assistants and is encouraged to talk through the process with the more experienced assistant. Hopefully, an experienced shop crew and production sound operator will be employed to help fill the gaps the new assistant needs help with. This system can be very effective but can also lead to frustration. It is like throwing someone into the deep end and hoping he or she learns to swim. Inevitably, mistakes will be made that can frustrate the shop and the crew, but if that person pulls it off and learns from it, then he or she is on the way to a very bright future.

A mistake that happens all too often is when a young designer tries to replicate this structure before he or she is ready. It is important to understand where you are in your career. A well-established designer has spent years building relationships with mixers and crew and assistants and theater house heads. An established designer can have his pick of sound operators. He has a stable of people who have worked with him repeatedly and who understand his likes and dislikes. He or she has the experience and rapport with the producer and technical director to point out when something is not going to work and possibly avert a crisis. This is not something that happens automatically when you get your first design job.

This structure works when you have highly experienced people working on the project. If you have talented and experienced people working with the assistant, they can help deal with any weaknesses the assistant may have; otherwise, it is a recipe for disaster. The mistake is that someone fresh out of college gets a job as an assistant for a big-name designer and is given a signal flow drawing and told to create the paperwork. That assistant does it, feels a great sense of accomplishment, and gets plenty of "atta-boys" for a job well done. Then that assistant gets an offer to design a low-budget Off-Broadway show. Now that the assistant is the designer, he decides to replicate the structure that he or she just experienced.

The first problem is that the new designer hires an assistant and expects that person to do what the new designer did as an assistant and create all of the paperwork for the system. Inevitably, there is less money available and the caliber of assistant and mixer is going to be lower. Something the new designer might not understand is how long it has taken for the established designer to develop his or her system. From naming convention to preference in the way a rack is built—this has developed over the course of years of designing. Inevitably, the new designer wants to position himself like the established designer, so he allows his team to do too much. The result can be a system that is labeled poorly and doesn't work properly, which can make tech a nightmare and be a huge black mark for a young designer's reputation.

The key as an assistant is to learn everything you can from the people around you. Learn the techniques of other assistants. Learn the techniques of the people building the racks. Learn why something is done one way and not another. And learn that if you want to be a designer, then you have to start developing your system and your style—and you have to learn how to teach it to others so you can start to build your own stable of people to work with.

Production Sound

The position of a true production sound is somewhat rare in theater. If there is a production sound position, then you can bet that it is a very large-scale production with a healthy budget. In hierarchy, the production sound would be on the same level as the associate, but with no artistic license. The production sound position is strictly as a sound system builder and installer. In lighting, there is an equivalent position called the production electrician. There is almost always a production electrician, but rarely is there a true production sound. The job of the production electrician is to take the lighting plot from the design team and flesh it out. The production electrician will do the cable order for the shop and figure out the logistics of building the system and loading it in. The production electrician usually does not work on the run of the show but rather works until the show opens and then moves on to the next show. If the show tours, then the production electrician will figure out how to manipulate the lighting design into a touring package. Basically, the production electrician takes care of all the nuts and bolts of the lighting design and leaves the design team to only deal with the artistic aspects.

It is much harder to separate artistic from system design when it comes to musical theater sound design. I designed a tour several years ago in which we loaded into the venue and started tech, and then found out there was a problem with the building power: it kept crashing the lighting system. We would get 15 minutes into tech and the whole lighting rig would start to do weird things and then shut down. After an afternoon of this, there was a meeting with the lighting design team, the director, the technical director, and the production electrician. The result was that the design team wished the production electrician good luck fixing it and went to dinner. No one batted an eye at this, because lighting has done an excellent job in establishing its hierarchy. No one would expect the lighting designer to grab a voltage meter and run backstage to figure out the problem and fix it. That's not his job. That is the job of the production electrician; if it doesn't get fixed, the blame would fall on the production electrician. This is not to say the lighting designer isn't expected to be involved and offer ideas to solve the problem, but that person is not expected to be the expert on the nuts and bolts of the system. This is because the lighting designer has done an exceptional job of shifting that responsibility onto the production electrician, and everyone accepts this hierarchy.

This would never fly with a musical theater sound designer. If the intercom system were down all afternoon, it would not reflect well on the sound designer to wish everyone luck in fixing it and then head out to dinner. Maybe it is because most musical theater sound designers have worked their way up from being mixers, so people expect them to be more hands-on. Or maybe it is because sound designers have a hard time walking away because they used to be mixers. The fact is that sound has not done as good of a job in separating the system from the artistic. The result is that it is rare to have a true production sound person on a show who is solely responsible for the system and has nothing to do with the run of the show or the artistic integrity of the show. There are definitely designers who are trying to change that, and progress has been made. Some of the more recent sound designers to join the ranks are much more artistic, and the system-minded and mixer-turned-designer models are slowly going away. As things change, the need for a production sound position is becoming more important. In another ten years we may be at the point where lighting is now and will have successfully separated the artistic from the system and solidified the need for a true production sound position on all shows.

So exactly what does a true production sound do? In a word . . . paperwork. This person fleshes out the designer's design and adds all the cable and parts and pieces to make the system work. The production sound would create the cable order and rack drawings. He would plan out speaker rigging and oversee the build process. He would work with the assistant, associate, and designer to make sure they have everything they need. When it comes time to tour a show, the production sound would get involved to alter the show into a touring system. In the absence of this position, this work gets split up and divided amongst the existing positions of associate, assistant, and A1. Sometimes the mixer, or A1, will be given the title of production sound, but this is not a true production sound because he has crossed the line between artistic and system by mixing the show.

The result of having a dedicated production sound person is that you can employ a less experienced assistant and A1 because they are not responsible for the nuts and bolts and do not need to have a full understanding of a sound system or loading in a sound system. It becomes the production sound's responsibility to know and understand the designer's likes and dislikes and his or her vision. It becomes the production sound's responsibility to staff the build with experienced people and take on the mundane minutia involved in putting a system together. It also frees up the A1 to concentrate on the mixing of the show rather than worrying about the buzz in the Spots channel on the Clear-Com. It can lead to an easier tech because the mixer is less likely to be burned out and overworked and is thus more able to mix a better show. It frees the mixer up to worry only about the artistic integrity of the designer's vision, which in the end is the most important aspect for the mixer.

A1/Head Sound

The role of the mixer, in its simplest definition, is the person who moves the faders based on what he or she hears to make a show sound balanced and pleasing to the audience while maintaining the aesthetic wishes of the designer. There are dozens of names for this person. A1 stands for Audio 1. (Lighting has an equivalent E1, E2, and E3, for Electrician 1, 2, and 3.) They can also be called *head sound*. This title can be a little deceiving. In regards to the union, all sound positions fall under the Electrics department and are therefore assistant electricians, but it is pretty common practice that sound is seen as its own department within the Electrics department. Therefore, you are the head of a subset of a department within a department. Just remember that you still answer to the house head electrician. Then there are times when you can be called *production sound*, which elevates you to being more responsible for the system build and installation, as discussed in the previous section. You can also be called the *sound mixer* or *sound board mixer*, a simple and self-explanatory title. Sometimes you are called the *sound operator* or *sound board operator*. This is the equivalent to the light board operator. This is another simple and self-explanatory title, and probably the best and most succinct title. Finally, at times you are called the *sound engineer*, which seems to elevate the position into something it really isn't. Engineering connotes some form of system component design that really is not part of the job. The term *sound engineer* comes from the recording studio world and was adopted by many in theater as a way to label the position of being the person in charge, but I have found most Broadway musical people call themselves *mixers*. Maybe it is because our real area of expertise is mixing. But no matter the label, this is the person who mixes the show.

By far the most important task for this person is to mix the show. If you can't build a show but you have great ears, then someone can be hired to build the show. If you can't load-in a show but you never miss a pick-up, then someone can be hired to load the show in. But if you can't mix, then you will be replaced. And just to be extremely clear, let me repeat: *If you can't mix, then you will be replaced.* The minute you decide you are irreplaceable is the minute when you start to lie to yourself. No matter how hard it is for the show, if you can't mix it is harder for the show to keep you around than to let you go. A bad mix can sink a show. Bad lighting is annoying. Ugly costumes are an eyesore. A bad set is disappointing. But bad sound is a reason to ask for your money back. In all of my years in the business, I have never heard of anyone asking for a refund because the lights were a little dark, but there are plenty of stories of people wanting their money back because of the sound.

I know of a theater that went into a panic after a musical tour came through that had disastrous sound. The theater had a large subscriber base and after the tour left literally hundreds of subscribers called to complain and cancel their subscriptions. This could have bankrupted the theater. If a subscriber-based theater loses its subscribers, it might as well close its doors. This led the theater to hire an acoustic consulting firm and pay them a very large sum of money to come in and fix the sound problems in the room. The firm moved the speakers, hung new ones, EQ'ed the room, hung acoustic panels, and did everything they could to make the room sound great. I have been there since the renovation, and I have to say they did a nice job. However, having mixed in that room many times before the renovation, I can attest to the space being hard but definitely not impossible. Even though the room was challenging, the real problem was the mixer of this tour. A poor mix almost collapsed a theater and in the end cost them hundreds of thousands of dollars to upgrade their system.

This is a scenario that has happened all over the country. As a touring mixer, it is common to walk into a house and be told the story of a show that sounded so bad that the powers-that-be wrote a big check to have someone come in and *fix* the room. Inevitably, the mixer is told that he will not need to bring in his sound system because their system has been meticulously tuned and the room is (this is one of my favorite quotes and I have heard it so many times over the years) "acoustically perfect." Then the battle begins as the A1 tries to explain why it is so important that he use his own gear. The key is for the A1 to keep a level head and explain that the system is tuned and that tying into their system is not as easy as it sounds. I have designed dozens of tours, and I tell all my mixers the same thing. Using a house system is a last resort. The problem is that you do not know what you are plugging into. Most of the time these systems are EQ'ed and the EQ is password-protected, which means you are stuck with it, whatever it is. I have known more than one mixer to be fired for using a house system and having a terrible show.

The problem is that it is not always about the gear. We all want good gear, and we can all get a bit snobbish about the superiority of this console over that one, or this mic over that, but the reality is that it comes down to the quality of the mixer. The designer can design the most amazing system with every bell and whistle, and that system can sound amazing, but if the mixer isn't good, then the show could sound mediocre at best. The fact is that one of the best shows I have ever heard was mixed on a Mackie (the Ford Escort of mixers) and one of the worst shows I have ever heard was mixed on a Cadac (the Jaguar of mixers). I have also heard two people mix the same show on the same system on different nights and heard completely different shows. A 2dB push on the sax in one line of a song might be exactly what makes that song perfect. That is not about the gear. That is about the mixer knowing what to do.

I have been sent out several times to fix tours that are having sound problems. Most of the time what I find is a good system with very nice gear that isn't being used properly. The result is a show filled with feedback and tinny distorted vocals. One tour in particular was a good example of this problem. When I arrived, I found that the gains on almost all of the wireless mic inputs were either maxed or almost maxed. When I PFL'ed a mic and talked into it the sound was immediately distorted and yet kind of quiet (PFL means *pre-fade listening*). The wireless was Sennheiser SK-50 packs with MKE-II mics going into a Midas Heritage H3000 console. It was all very good gear. So why was the gain so high?

The answer is EQ. This is what I consider a rookie mistake, and I have seen it so many times. A mixer starts to EQ a mic and makes some cuts. Sometimes they pull an old trick where they put all the mics onstage and turn them up until they feedback and start cutting EQ. (I am just going to say that this is a horrible idea, and if you ever find yourself with an urge to do this, please take a moment and stop yourself. I promise you that something else is wrong with your system.) As he makes cuts in the EQ, the mixer has a problem. He is losing gain. So gain up. New problem. The bad frequencies are back and now there are new ones. So more EQ. After a while you end up with a mess. On this show, the gains were almost maxed and the EQs were almost all completely cut. Then, when there was no more EQ on the board to cut, the mixer inserted an EQ on the vocal mix and continued to cut. When he ran out of places to cut on that EQ, he actually put a second EQ inline and started cutting more. The result was great gear and a horrible sounding show.

This is not to say that if you can mix then you can sit back and relax because you have it made. The best A1s in the business are well-rounded. They can mix and build and load-in. They are good managers and are detail-oriented. The best mixers can walk up to just about any piece of gear and figure it out, and they can throw faders effortlessly. The best mixers are constantly looking to learn the latest gear and are extremely protective of their ears. The best mixers are like the best artists. They understand that their value and their art is crafting an emotional mix, but in order to do that they have to understand and master the technical and political aspects of the business. Picasso didn't start by painting Cubist painting. He learned how to use his tools and then learned how to manipulate them.

The sound board operator is also the catch-all for the system. It falls on the A1 to catch all of the things that have slipped through the cracks and make sure the system works the way it is supposed to. The A1 has to make sure the intercom system is tested and working. He also has to make sure the video system and paging system are working. He has to make sure the main PA is working, and he has to run a crew.

Typically, the design team will give notes to the head sound, and then it is his job to delegate who will do the notes and make sure they get done. It can be an incredibly demanding and exhausting job. The A1 works from 8 am to 12 am six days a week for weeks to get a show open, and then once the show is open the A1 isn't allowed to take the day off no matter what. The A1 is the only person who knows how to mix a show, and if the A1 didn't show up, the show would be very quiet. Broadway is littered with stories of sound people throwing up while mixing because they are so sick and don't have a sub trained. It is a hard job, but it is also incredibly rewarding and people do it because they are passionate about it. Mixers love what they do. When they stop loving it, they get out, because it is too hard for anyone who doesn't get some kind of joy out of standing behind a console and being part of the spontaneous artistic creation.

A2/Deck Sound

> The A2 position can also be called the *deck sound* position and sometimes a *mic tech*. There are shows in which there is more than one sound position on deck, in which case there could be an A3, A4, A5, etc. Typically, these positions would just be called deck sound positions. If the show is a tour, it usually travels with an A2 and the local sound person is called a deck sound. If the show does not travel with an A2, then the local sound person is still called a deck sound. The deck sound person is responsible for everything from the proscenium line to the back wall. His job is to prep the mics, manage the inventory, maintain the mics, deliver the mics, mic the actors, do the cues in the show, and check for mic problems. It is quite a lot of responsibility.

Deck sound people are usually more technically minded than conceptually minded. There are some deck sound people who also mix, but it is not a requirement. Every musical on Broadway has a sub mixer, which is someone who knows how to mix the show and mixes at least one show a week to stay fresh on the show. The job of the sub is to be there in case the main mixer is not able to do the show. It has become common on Broadway to hire a deck sound person who can also be the sub on the mix so that there are always two people in the building who can mix the show. Having a good deck sound person is extremely important to a musical. A good deck sound person will do preventative maintenance to make mics last longer and will find a bad connector and take it out of the system and replace it without the mixer ever knowing it happened.

The main task of the deck sound person is to deal with the wireless microphones. The first part of this is to understand the designer's goals in mic'ing the show. Some designers are adamant that the mics never be seen. There are some producers and directors who insist the mics never be seen. Then there are other designers who don't mind seeing the mics and are only worried about the sound. There could also be mics on headbands or booms. The revival of *A Chorus Line* on Broadway was a great example of hiding mics. None of the creatives, which is the term used for the group of directors and designers on a show, wanted to see a mic or a mic wire or a mic pack. Most of the cast was in leotards with low-cut backs. The first problem was where to put the mic packs. The deck sound person worked with the wardrobe department and had pouches built into the bras for the women and the dance belts for the men. The next problem was where to put the mic. For some women it was not possible to run the mic to their head without seeing the cable so those mics ended up in the women's cleavage. Other mics were fished through elastic loops, sewn in by wardrobe, to get the mics to the shoulder and then up the neck under the hair and to the forehead. In the end, the effect was stunning. The audience could not see any clue that the actor was wearing a mic and this was all due to the deck sound person. *Rent* was a very different show. The creatives on that show actually liked seeing the mics and the whole cast was placed on booms, which gave the designer, Steve Kennedy, a lot more headroom or gain before feedback.

Once the deck sound person knows what the goals are, then that person can start prepping the mics to meet the goals. The first step is labeling. The deck sound person will label every receiver and transmitter with a number. Then a separate label will be added with the track name for a cast member as well as his or her real name. These are separate labels because the number relates to the hardware, but the cast member might shuffle through different packs throughout the course of the run of the show. Next, a label with the actor's name is attached to the mic near the connector. This is usually a tiny flag label. Next, the cable near the connector is doubled over and a Hellerman sleeve, which is a 1" rubber tube, is used to attach the cable to the connector. This creates a strain relief for the connector, which is the most fragile part of the mic. Next, a cap is placed on the mic, which could either be flat or high boost depending on the designer's needs, and usually a piece of waterproof tape, such as Elastoplast, is wrapped around the cap to deter sweat-outs.

Certain tools are very important to a deck sound person. One of the most important tools is the Hellerman tool, which is used to spread a Hellerman sleeve open. This tool can be used to strain relief connectors or attach mics to booms or ear loops. Ear loops are another important tool. These can be made of plastic, metal, or rubber. Ear loops wrap around the actor's ear to hold the mic in place. Toupee clips are also important. They are used to hold the mics in the hair and come in different sizes and colors. It is good to try to match the toupee clip with the hair color. Toupee clips are prepped by tying elastic string across the clip; the mic slides through the string. Usually three toupee clips are used. One is very close to the head of the mic. The second is near the crown of the actor's head. The third is near the hairline on the back of the neck. Artist craft wire is another important item. It is used to stiffen a mic so it can be bent into the position needed. Usually craft wire is wrapped several times around the mic near the capsule. That allows for the mic to be positioned and to maintain its position. Sometimes a Hellerman sleeve is placed on top of the craft wire to help it hide better and be smoother. Figures 3.1 through 3.4 show some example pictures of some mic rigging techniques.

When it comes to hiding the mics, the deck sound person colors the mic cable to match the color of the actor's hair. The mics can be colored several different ways. Sometimes spray paint can be used, but more commonly paint

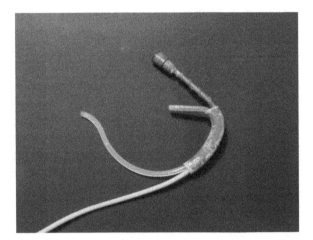

Figure 3.1 This photo shows a mic attached to a plastic ear loop using a Hellerman sleeve. The Hellerman has been colored (but needs a touch-up) to match the actor's hair. Craft wire is wrapped around the mic to stiffen it.

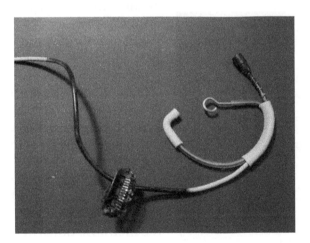

Figure 3.2 A mic attached to a metal ear loop. It also has a toupee clip to attach the mic to the actor's hair.

markers such as Pentel paint markers are used. Then, using the toupee clips and a rat tail comb, the deck sound person clips the mic in the hair and hides it as much as possible. If the actor is wearing a wig, the mic ends up going under the wig cap.

Sometimes it is possible to put the mic pack in the wig as well. Putting mic packs in the wigs is great for the life of the mic because they have much less wear and tear.

Once the mics are on the actors and hidden, the deck sound person's job is to sit by the RF rack and monitor the RF. This entails listening to the show feed for issues and

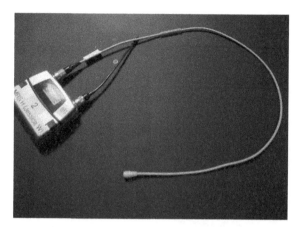

Figure 3.3 An 18" mic made short for placing in wigs.

Figure 3.4 This shows the normal method of wrapping the mic to prepare it for delivery to the actors.

using headphones to listen to individual mics to find problems when the actors are offstage and fix the problem before it gets onstage. The deck sound person sometimes monitors the wireless mics using a computer hooked up to the receivers. If a problem is found, such as a bad and crackling connector, the deck sound person has to decide how to deal with that problem. Usually the first response is to disconnect the receiver from the system by unpatching the XLR or by turning off the receiver. Then it is the deck sound person's job to know the show well enough to know if that actor has any upcoming lines or is offstage for a while. The deck sound person then has to find the actor and replace the mic.

The deck sound person is also responsible for cues back stage. These could be cues for mic swaps or cues to pat sweat off of an actor and spray out the mic with air. The Broadway production of *The Full Monty* had a very important deck sound cue. The final scene in the show is the scene where all the men strip down to nothing. Literally nothing. So before that scene the men had a quick change to get them out of costume and mic and into their police/stripper outfit. So where do you put a mic and pack on a nude man? You hide them in their hats, of course. The deck sound person had a cue to prep mics in the police hats and then be at the quick change to make sure to get everyone's mics as the quick change finished.

The deck sound person is also responsible for maintaining the mics. This involves touching up paint periodically and replacing toupee clips as they age and lose their snap. It also involves re-applying Elastoplast to the caps and cleaning the tape goo off the mic cables. It also involves an ultrasonic jewelry cleaner. The caps on the mics can become caked with hairspray or make-up and they have to be cleaned, and the only way to clean them is to use an ultrasonic jewelry cleaner. If there are handhelds in the show, they also need to be cleaned periodically, or they begin to smell. To clean these mics, the deck sound person takes some mouthwash and a toothbrush and scrubs the outer portion of the mic head.

Finally, on a daily basis it is the deck sound person's job to battery up the mics. The battery-up process actually has its own prep. If the show is using conventional alkaline batteries, then each battery is tested with a VU meter. This usually happens the day before, but sometimes the deck sound person will check all the batteries needed for the week and have them ready. If the show uses rechargeables, which is becoming more popular, the batteries have to be conditioned after so many uses according to the manufacturer specs, and they still have to be metered. Once the mics are batteried, the deck sound person checks the In/Out sheet from stage management that explains if there are any cast changes in the show and adjusts the mics accordingly. It is possible that cast changes for a show could add a cue for the deck sound. Sometimes one actor will cover another actor's track for one scene and the deck sound will have to manually repatch the wireless backstage so the mixer out front does not have to change the mix of the show. Once all of this is done, the mics are tested through the system with the A1 and then they are delivered. There are several delivery styles. Some deck sound people like little plastic buckets and some like shoe bags. It is the debate for the ages.

As you can see, the job of the A2 or deck sound can be very complex and very hectic. Finding a good A2 is crucial to a show running smoothly and with few problems. This is why there are deck sound people on Broadway who have worked in the same theater for years. Once a theater finds a good A2, they hold on to that person. Some important A2s on Broadway include Dave Shepp, who works at the August Wilson Theater; Bonnie Runk, who worked at the Al Hirschfeld for years; Bob Beemers at the Broadway; John Cooper at the

Imperial; Steve Carey, who has been the A2 on *The Phantom of the Opera* for 20 years; and Randy Morrison, who works at the Shubert. Randy Morrison is probably the most famous deck sound person on Broadway because he also fixes broken connectors on mics, so everyone knows Randy.

Monitor Mixer

More and more shows have started to add a *monitor mixer*. This is the backstage equivalent of the A1, with the biggest difference being his audience. The A1 mixes for the audience in the house, while the monitor mixer mixes for the cast and musicians. The skill set is basically the same, but there is an added level of politics. The A1 has to deal with notes from the creatives, such as the director or composer, when they watch the show, and the A1 has to deal with complaints from the audience in the house. On the other hand, the monitor mixer gets notes constantly from the cast as well as stage management and has to find a way to make everyone happy.

The rise of the monitor mixer position is happening because of the change in the type of shows being created. In rock and roll sound, the monitor mixer is a given. No one would even question the need for a monitor mixer, but in theater it has not been so common. As shows have begun to increase in complexity and as the music in modern musicals trends toward more modern rock and roll, the necessity for a monitor mixer has been increasing. Also, as celebrities who are used to wearing in-ear monitors continue to cross over into Broadway, they bring with them an expectation of what it should sound like onstage.

When a show does not have a monitor mixer, this lack of personnel can be used effectively to curb the impossible requests that sometimes come up. As the cast starts to request more and more monitors onstage, at a certain point the designer can go to the producer and director and tell them that it can only be done if they hire a monitor mixer. If the show can't afford that, then the producer and director will usually, but not always, work to help calm the cast. But if the show can afford a monitor mixer, then there is nothing better. It is very liberating for an A1 to be able to only worry about the sound in the house. It is so much less stressful to know that someone else is taking care of the cast and musicians. After that, the only problem is for the A1 and monitor mixer to work together to make sure the monitor levels are not affecting what the audience is hearing, which can be a battle unto itself.

DIFFERENT LEVELS OF THEATRICAL SOUND

The next important thing to discuss is the different levels of theatrical sound. No one starts off as a mixer on Broadway. Everyone started somewhere and worked his or her way up. The crucial point I would like to make is that it doesn't matter what level you are at or what the budget is: the methods and goals of mixing musical theater are always the same. We are there to help shape the emotional arc of the show. Inevitably, you will have better gear on a Broadway show than on an Off-Off Broadway show, but that doesn't mean you treat the product any differently and it doesn't mean you can't still have an amazing show. One of my all-time favorite moments as a designer was working with a mixer in a small 150-seat theater on a musical. We had some beat-up old Ramsa speakers for the mains and about 30 JBL Control Ones for delays and surrounds, and the board was a Mackie. There was one moment in the show that just gave me chills, and that moment remains more impressive to me than some of the moments I have heard on million-dollar sound systems.

As a mixer, it is important to set goals for yourself and to know that in every situation you can learn something. The lesson may not be clear for months or years to come, but if you look out for it, you will find the lesson. Also, it is important to learn the techniques used at the levels above yours and adapt those techniques as best you can to where you are. If you learn mic'ing and mixing techniques, you can add them to your bag of tricks. In the end, that is what it is all about—building a bag of tricks so that when you run into a situation, you can pull something out and have a solution to a problem.

As you grow as a mixer, you have to set goals. There is a path to success as a mixer. If you know what level you want to get to, then you set your sights on that goal and know that there is a road that will lead you to that goal. One advantage we still have as sound people is that there simply aren't enough of us out there. If you put yourself out there and work hard, it will be noticed; you will be snagged by someone and you will move up. A big key to this is staying open to knowledge. Keep in mind that people above you are there for a reason and that they probably know something you can learn. Don't ever close yourself off to that. If you are convinced that you are the smartest person in the room, then you are done growing and you will surely miss out. Mixing is a skill that is passed on from generation to generation. Stay open to it.

Small-Scale and Off-Off Broadway

So you got a job mixing a musical for a 50-seat black box theater and are getting paid less than the barista at your local Starbucks, and yet you are excited about the opportunity. Good for you, and congratulations. Welcome to the business. This is a great place to get started. It usually requires a younger person who is more interested in the art than the money, and it is usually a great community of people that can quickly start to feel like a second family. There is nothing like eating, sleeping, and breathing theater.

The pitfalls of this level of theater really come down to budget. There is never enough money at this level to get the top-notch gear. The challenge is to learn how to do amazing things with nothing. The beauty of this level is that it is almost always about passion. From the producer to the director to the designer, people are working on a project because they are excited about the project. Sometimes they are doing it just because they love the process. The results can be truly awe-inspiring.

The warning at this level is the same warning I feel at all levels, but it is a good one to learn right off the bat. The question is, what is the difference between a good sound mixer and a bad sound mixer? The good sound mixer knows what jobs not to take. There are times when you have to say no, and there are times when you have to walk away. You have to learn to size up a job and know whether it is a disaster in the making. If you take a job that is destined for failure, then that black mark is on you. If you take a show where they want you to mix a rock musical with a live rock band onstage using area mics for the vocals and you take the job, then you will be remembered as the person who ruined their show.

As hard as it may be, there are times when you have to walk away. I have turned down many jobs because I didn't think I could pull it off. The result is that the jobs I have done have mostly been successful. Of course, there are some that I should've turned down but didn't, and there are some that I couldn't turn down because I needed the money. You do the best you can. Just know that sometimes walking away is the best thing you can do for your career.

If you have to do the job, then the next lesson is CYA: Cover Your A . . . It is acceptable to respectfully voice your concerns. There are times at this level when there is no sound designer. Sometimes you are mixer, designer, A2, and usher. One of my all-time favorite job postings actually read, "Looking for Sound Mixer. Duties include: setting up sound system, mixing shows, and cleaning the restrooms." Needless to say, I did not apply. When this is the case, you are the expert. The company actually needs you to let them know if something isn't going to work. They may not be able to fix the problem and they may not accept it, but you have to tell them. Otherwise you will end up mixing a show with a rock band onstage in a 50-seat house with two shotguns pointed at the stage for vocals and then getting chewed out by the director because he couldn't hear the lyrics. If you pointed out the problems in the beginning, then you have a leg to stand on. If you didn't, then you have lost all credibility and you will either be replaced or never work for that theater again. Believe it or not, even at this level, where there is no budget and the pay is terrible, you can still get fired for mixing a bad show. Sound is just that important.

Now for a quick rant about area mic'ing. Unfortunately, it is very common for shows to use area mic'ing. Area mic'ing is when you take some PCC-160s, or boundary microphones, and some shotgun mics and place them around the stage, and that is all you use to amplify the show. The result usually ends up being disappointing. Area mics are not meant to amplify more than about 10dB over the acoustic level and can only cover about 8 feet per mic, so you need lots of area mics, and the more mics, the more likely you will have feedback. There is a common misconception among non-sound people that you can hang a mic in a room and that mic can discriminate between all of the sound in the room and the sound of one voice and just amplify that one voice. It is hard to explain to people that it just doesn't work that way. It is hard to convince them to give up on the area mic'ing idea, but if at all possible, try to talk them out of it. If you are lucky, you have built enough of a relationship with the theater that they will trust your judgment; if not, then good luck.

Regional Theater and Off-Broadway

Once you have stepped up to this level, you will start to see better gear and probably bigger designers. It is becoming more and more common for Broadway sound designers to design shows at regional theaters around the country. This has been the case for other design disciplines for decades, but it used to be that sound design for regional theater was handled locally. As sound has grown to a more accepted and respected design field, the regional theaters have stepped up their expectations. The positive benefit for mixers at the regional theater level is that they now get exposure to more people. This makes it easier to make a move to the next level, if that is your goal.

Even though the budgets are bigger at regional theaters, there are still money problems. When dealing with a theater that has a budget of $100, there is usually an understanding between all involved that whatever they get is the best they can hope for with that kind of budget. Of course, there are exceptions where they expect more, but usually they understand that there are things they can't afford. The result is that they are more accepting of things like the thunder not rattling the house like they hoped, because they couldn't afford the subs. But things are different at the regional theater level. The problem here is that they actually spent real money on a sound system. It is not uncommon for a regional theater to spend a large chunk of money on their sound system, and when they do, their expectations rise dramatically.

After a theater spends $10,000 on 30 wireless transmitters, receivers, and lavalieres, they feel like they should have shows with no wireless mic problems. There are even times when they will come right out and say, "We just spent $10,000 on these wireless mics. Why are they crackling? Why are they sweating out? Why are they breaking?" The reality is that if you bought top-of-the-line wireless gear, then those 30 mics would cost between $150,000 and $200,000, but the last thing the theater wants to hear is that they bought cheap gear that is going to cause more problems than it will fix. The most important lesson to learn from this is that if you find yourself working at a regional theater and they want to spend money on their sound system, make sure you guide them down the path of buying less gear that is of higher quality. Otherwise, it is going to reflect on you when the cheaper gear doesn't work.

When it comes to Off-Broadway, it is similar to regional theater in the pay structure and budgets. The main difference is that regional theater is not-for-profit, whereas Off-Broadway is for-profit. This difference is huge in the way the two produce shows. The regional theater has donors and receives grants for things such as the children's theater program. The regional theater also has a subscriber base, which means that they can sell tickets a year in advance for their shows. The regional theater builds a reputation over time and a subscription base by consistently providing a certain level of entertainment. The Off-Broadway show has investors and doesn't start selling tickets until right before the show opens. There is no subscriber base and patrons buy tickets based on whether they want to see a particular show instead of knowing that the show will probably be good based on past experience with the company.

The Off-Broadway show usually does not own its own gear and usually rents a theater that is a "four-wall theater," which means a theater with no existing equipment. The show has to supply everything, which means the Off-Broadway show will rent gear from a sound shop and load it in. Off-Broadway, which by definition is a theater in Manhattan that is between 100 and 499 seats, is treated almost identically to Broadway, which makes it great training ground for a mixer who wants to mix on Broadway. The designers are usually Broadway designers and there is always a chance that your Off-Broadway show will move to Broadway. Examples of this are *Avenue Q*, *Urinetown*, *Bloody, Bloody Andrew Jackson*, and *Noise/Funk*. Regional theaters also transfer shows to Broadway on a regular basis. Some examples include *The Full Monty*, *Dirty Rotten Scoundrels*, and *Wonderland*.

Off-Broadway being for-profit leads to very different ways of spending. The regional theater is usually working on a budget that could be scheduled for years, whereas the Off-Broadway theater has an immediate budget to get the show open and keep it running. The regional theater is more likely to buy equipment that can be used for a decade, while the Off-Broadway would prefer to rent the equipment and have the shop take care of replacing gear as it breaks. In some ways, the Off-Broadway show is easier to deal with because they are willing to buy and rent more expensive equipment, as they are not worried about the future as much as how the show sounds on opening night when the critics are in the audience. But, on the other hand, the Off-Broadway show is much less accepting of sound problems. For an Off-Broadway show, one bad review could close the show and the investors could lose their money, while a regional theater can absorb a bad review without bankrupting the theater.

In the end, this level of theater is a great training ground for anyone who wants to mix on Broadway. There are some amazing sound department heads around the country who do a great job training young mixers, and there are some great people who work Off-Broadway who are highly skilled and have a wealth of information. Absorb as much as you can at this level, and if you find that you are happy mixing and working at this level, then you have found your career. This level of theater is typically the highest level of truly risky and artistic theater. This is where you get to mix that show that isn't commercial enough for Broadway but blows your mind. Mixing and working at a regional theater can be highly fulfilling, and mixing Off-Broadway can truly be exhilarating.

Touring Theater

The next level just under Broadway mixing is touring mixing. There is a definite life cycle to a show. A show starts at a regional theater and then transfers to Broadway. It runs for at least a year and then that show is remounted as a touring production. The first national tour of a Broadway show usually moves very slowly. When *Wicked* first started touring, it would sit in a city for weeks or months at a time. *The Phantom of the Opera* holds the distinction of being the longest touring show in US history, and up until the day that tour closed, its shortest stop was a couple of weeks. After most shows tour for a while, the show starts to shorten its time in each city. At first a show goes to what are called "A" cities. These cities have a large enough population to sell tickets for long periods of time for a show, but after a show has been to all the "A" cities, it starts going to the "B" cities. "B" cities can usually only sell an eight-show week's worth of tickets. Next, a tour starts hitting smaller cities and moves anywhere from twice a week to every day. Once a tour hits this level, it is called a "bus and truck one-nighter." This cycle usually takes several years, and after the bus-and-truck tour the show closes for good.

As a tour moves through its life cycle, the show has to adapt to each level. At first it is possible to basically move a Broadway show from city to city. A load-in for a first national tour could be anywhere from two days to two weeks. *Phantom*'s load-in took almost ten days. But as the show starts moving more, the load-ins have to get easier and take less time. This is accomplished by making changes to the equipment on the show to accommodate quicker load-ins. Also, the size of the show starts to shrink. A first national can take dozens of tractor trailers to move. *Phantom* took twenty-six 53' trucks to move it. As a tour moves toward doing one-nighters, the tour has to find a way to fit in fewer trucks. This helps make load-in quicker and saves tons of money on trucking.

The mixers who are mixing the first national tours are typically mixers at the top of their game. These are mixers who have toured for years and maybe even mixed on Broadway. These are well-paying jobs that are hard to get. The only way to get a mixing job on a touring Broadway musical is to be picked by the sound designer, which means you have to build a relationship with designers so they know they can trust you to tour their show. As a tour heads toward the one-nighter circuit, the pay decreases and the experience level is lower. At the one-nighter level many of the mixers are straight out of college or from a regional theater. Most of the mixers on the first-national tours started out by doing a one-nighter tour and worked their way up.

If your goal is to mix on Broadway, then doing a tour is almost essential. The knowledge you gain from touring just can't be gained anywhere else. The budgets for tours are sometimes very close to budgets for a Broadway show, and the equipment used on a tour is very similar, if not identical, to what is used on Broadway. Being a mixer requires you to know and understand a wide variety of equipment. Some of this equipment is so specialized and expensive that it just isn't used outside of Broadway or touring. For that reason, if you want to clock some hours on a Midas XL8, then your best chance outside of Broadway is to find a touring position.

A lot of Broadway mixers were touring mixers, and a lot of Broadway shows start somewhere other than New York. These shows have an out-of-town tryout to work out the bugs and tighten the story. In order for a mixer to pull this off, he will need to have an understanding of touring a show and how to load-in a show. There is no better training ground for mixers than touring. You learn the nuts and bolts and you learn to mix a Broadway-scale musical with Broadway techniques. You also learn how to EQ and optimize a system. A bus-and-truck one-nighter tour can play 100 cities in 40 weeks. That is 100 venues that have to be EQ'ed. After you do that, you will have a strong understanding of how to EQ and you will be a pro at loading-in and out.

If you decide that you are interested in touring, there are several things you should know. The first is that it is hard work. Don't let anyone ever tell you it isn't, and don't do it if you are looking for an easy job. At the same time, it can be incredibly fun. The normal schedule for a bus-and-truck one-nighter is an 8 am load-in followed by a sound check at 6 pm and a show at 8 pm. If it is a one-nighter, then there is a load-out after the show. The day is over after the load-out, which will be done by 2 am. Then you climb on a bus and sleep while you are driven to the next city. If it is not a one-nighter, then you are off to the hotel.

On a bus-and-truck one-nighter, you spend most of your time living on a bus. Bus living is great. The bus is split into three lounges. The front lounge is like a living room. There is a couch on each side and a TV in front. There is a little kitchenette with a sink and refrigerator and a little dining table. There is also a bathroom that consists of a toilet and a sink. The middle lounge is the sleeping lounge. The sleeping lounge has six bunks on each side of the bus and a hall down the middle. There is a bottom, middle, and top bunk on the passenger side front and rear as well as the driver side. The bunks are not large but they are comfortable. A curtain is at every bunk and can be closed for privacy. A lot of the bunks now have 5.6" LCD TVs for watching movies. The back lounge is like a second living room. It usually has a wrap-around couch and a TV. There are bays under the bus to keep luggage.

If you have decided that you want to tour, then more than likely you will get a job on one of the smaller bus-and-truck tours. The first tour you do probably will not pay all that well, but that is acceptable if you just keep in mind that you are gaining the knowledge you need to move to the next level. You will most definitely work hard and you will learn more than you ever wanted to about being a stagehand. In the end, that is what a mixer is—just a stagehand that specializes in mixing sound. If you think it is all about mixing, you are in for a surprise. Even though the bus-and-truck schedule sounds grueling, people have been doing it for decades. It is hard, but it can be incredibly enjoyable. As you move up to bigger and better tours, you will find that the work never gets easier.

I had a mixer who did a bus-and-truck tour for me and he did a great job. The following year, he turned down an A1 position to take an A2 position on a bigger tour because he thought the tour sounded easier. There was a two-day load-in and the show was going to sit more often. I laughed and told him it doesn't get easier, just bigger. The schedule on a tour like that is an eight-show week with two shows on Sunday followed by a load-out. The load-out can take anywhere from 8 to 12 hours, so you load-out until around 8 am. Don't worry about eating. You will usually have a catered meal sometime that night. After load-out, you sometimes go straight to the airport and fly to the next city, and sometimes after you land you check in at the hotel and then go to the theater for a four-hour pre-hang, which is the part of the load-in where motors and truss are hung, among other things. My friend went from a grueling one-nighter schedule to an arguably more grueling sit-down schedule.

The whole point is that touring is the best thing you can do to learn your craft. And mixing sound is a craft. There is an art to it, but if you don't know the nuts and bolts, you are only going to go so far. If you really want to succeed, you have to be a good stagehand first and a great mixer second. You can learn a lot about being a good stagehand from working in theaters in college or regionally or other places, but touring is an incredible experience that can't be duplicated in any other way.

The toolkit for a touring sound person should include a crescent wrench, snips, a soldering iron, wire strippers, a multi-tool such as a Leatherman or a Gerber, a set of jeweler's screwdrivers, a set of screwdrivers, a socket set with ratchet, gloves, a multi-meter for testing voltage, a Q-Box, a cable tester (that can test XLR, BNC, NL4, NL8, RJ-45), a P-Touch, and a cordless drill.

Broadway Theater

So this is the goal. Broadway is the crown jewel of musical theater. If you are a sound mixer and you mix musicals, then the highest level of mixing is Broadway. That is not to say that the best mixers are on Broadway or that the best sounding shows are on Broadway, but that the highest level of talent, quality, budget, and scrutiny is in the heart of New York City, beating away in fewer than 40 theaters residing in a 15-block radius. People come from all over the world to see a musical on Broadway. The crews are top notch and the shows are impressive spectacles.

Broadway theater is very different from other levels of theater. There are unique practices and methods. There are special rules not found in other places. There are different unions and contract issues. There are budget issues and safety issues that are different. It is a different world, and this is what we will break down for the rest of this book.

Hierarchy and Loyalty

It has been mentioned before but is worth discussing again. If you want to have a long and lucrative career in this business, you have to learn your place. You need to understand where you are in the pecking order and what your responsibilities are. More importantly, you need to understand what you are not supposed to do or say. You have to be careful not to overstep your boundaries. Sometimes the best answer is, "Let me check with the designer and I will get back to you." You have to be careful not to commit to something that later the designer has to un-commit to. Some designers are laid back when it comes to things like this, but it is more common for designers to be very protective of communications between the sound team and the rest of the company. It would be a very long day if you, as the mixer, told the lead, "I would be glad to put your vocals in the monitors onstage," because you would've just dug yourself into a hole that you may not get out of.

A designer is looking for a mixer who can mix a show, load-in a show, build a show, plan the build, and make the designer look good. Designers want the whole package. They want someone with a great ear who is humble enough to take notes. They want someone who can "protect" their design. That word is used a lot because the design is always under attack. You constantly get notes from cast, musicians, stage management, creatives, or producers about what they think should be louder or softer in the show. You have to know how to respond. If you take every note and follow it, then the sound of the show could drift from what the designer wants. Then you will have to explain to the designer why the show sounds the way it does. Of course, the most common way around this is to smile and nod and say you will fix it and then change nothing. Amazingly, people will usually hear the show and think you fixed it. You just never know where their notes are coming from. Are they sick? Did they fly recently? Did they have a big meal or a glass of wine just before coming to the theater? A little-known fact is that digestion affects your hearing. While you digest food, some blood leaves your head and goes to your stomach. This causes a loss of high-end frequencies and general SPL (volume).

You have to be confident enough to know what the designer wants the sound of the show to be, and you have to protect it. Another acceptable answer is to say, "I am not allowed to change that without talking to the designer." This will usually result in that person talking to the designer, and if the designer agrees with the note, then the designer will call you and give it to you. I have worked on many shows where the designer lets the entire company know that no one is allowed to talk to the mixer about the sound of the show and that all notes go directly to the designer. It doesn't always work, but sometimes it does.

You are a reflection on the designer. The way you look and handle yourself directly reflects on the designer. If you are rude, then it will get back to the designer and he or she will be less likely to hire you again. The way you dress is also important. As a mixer you do your job in the house with the audience. If you wear grungy t-shirts and torn jeans, you cheapen the Broadway experience for the audience and you show disrespect for your profession. Some designers are very particular about the way their mixers dress. Some may request that you wear a coat. There was one mixer on Broadway who was wearing grungy t-shirts and show management approached him and asked him if he could dress a little nicer out front. His response was, "That's not in my contract. You can't tell me how to dress." This guy could be the greatest mixer in the world and the best production guy, but the fact is designers wouldn't want someone who would say that to mix for them.

Loyalty is another important aspect to this business. If you do a good job for a designer, then that person will show some loyalty to you. The designer will do what he can to get you another show or more work. In turn, if you are working with a designer who has given you work, then you should give him some loyalty as well. You do not take a job from a different designer without checking with your designer first, and you should make sure your designer knows and approves before you make a huge commitment. For all you know, the designer has a show for you that he just can't talk about yet. In the end, what is important about loyalty is that it defines the relationship between a mixer and a designer. Once a designer finds a mixer he likes, he will try to hold onto that person, and the only way to do that is to keep finding work for the mixer. As a mixer, the unspoken truth is that the designer's most important job is to find more shows to design. A mixer's job, in the end, is to take care of and do as much as you can for the designer so he can be free to find more projects. A good designer is never working on just one show. If the mixer does his job, then the designer can juggle multiple shows, and the more work the designer has, the more work the mixer has. It is a great symbiosis.

THE PREP

5

DESIGN TEAM'S JOB

All Broadway shows start with the prep period. This begins weeks, if not months, before the show and consists of the planning of and creating the drawings and paperwork for the system. It includes production meetings and site surveys of the theater. It is entirely possible that the mixer has not been hired for the beginning stages of the prep period. The prep period leads to the shop build, which is when the sound system will be pulled from a sound rental shop and labeled, built, and tested in the shop. The prep concludes when the system is loaded onto a truck at the shop and sent to the theater for the take-in or the load-in.

The job of the design team is to design the system and gather the information from other departments about their needs. Once the designer has the information he needs, he draws out the sound system. This could sometimes be as simple as a hand drawing on a piece of paper or a more elaborate Vectorworks or Stardraw drawing. The designer's job is to give a detailed overview of the system so his team can flesh out the system with rack drawings and cable schemes. Figures 5.1 through 5.3 are examples of the drawings that the designer will normally pass on to his team. It is important to note that USITT, the United States Institute for Theatre Technology, Inc., is working on standardizing sound drawings and terminology, but this has not happened yet and has definitely not made any headway on Broadway. These are drawings that you might receive from a Broadway designer, but they may vary from what USITT calls standard. I look forward to their standardization, but I think it will take years before we see it really take hold.

As you can see in Figures 5.1– 5.4, the point is to give a decent overview of the sound system. The drawings are simple but contain most of the information needed to flesh out the sound system. From these drawings, typically, the associate or the assistant will go a step further and start doing drawings that further explain the sound system. The next step would be to lay out some other basic information about the system. The most important aspect of any

Figure 5.1 A typical signal flow output drawing from a sound designer. USITT calls signal flow drawings *block diagrams*.

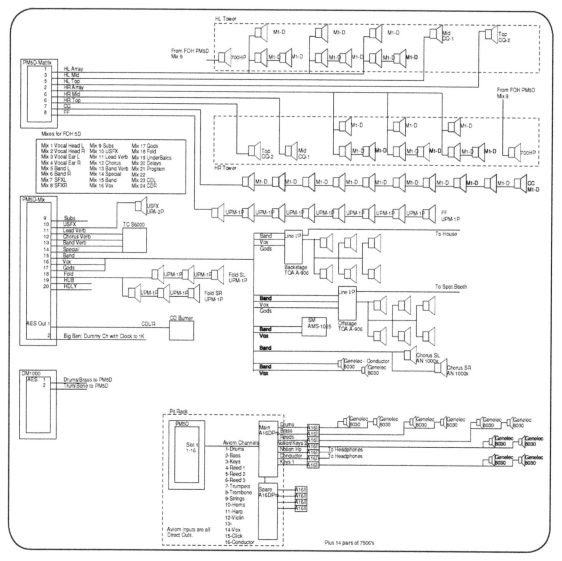

sound system outside of the inputs and outputs would be the intercom, video communication, and paging parts of the system. Inputs and outputs are relatively straightforward and easy to understand, but these other parts of the system can be very complex, and each designer has his or her own special method of doing these systems. It is up to the assistant or associate to lay these systems out in a way the designer likes, allowing for any special needs of the other design teams. Lighting has the most important requests when it comes to intercom. Some lighting designers like their com a very specific way, and their assistant is usually a good source to ask about the needs of the lighting department. Once this information has been pulled together, the associate or assistant will add more signal flow drawings.

The most important thing to note about these drawings is that they are overviews and not overly detailed. A mistake some people make is trying to overload information on drawings. Keeping drawings simple and to the point can be more effective. The idea of these signal flow drawings is that someone can pick up a drawing and quickly understand the components of the system. The idea is not to convey every detail about the components. These drawings rarely detail anything about cabling or specific focus or rigging. That information is saved for other places. Other drawings that can be helpful at this stage are signal flow drawings for the computers, midi routing, and the paging system. It is possible that the design team will

Figure 5.2 A typical signal flow input drawing from a sound designer.

DM1000V2		
Kick	M88 N(C)	1
Snare Bottom	SM57	2
Snare Top	SM57	3
Hi-Hat	4023	4
Tom Lo	E604	5
Tom Mid	E604	6
Tom Hi	E604	7
OHL	KM184	8
OHR	KM184	9
Glock	KM184	10
Timps	U89	11
Xylophone	4022	12
Toys	KM184	13
Trumpet 1	U89	14
Trumpet 2	U89	15
Trombone	U89	16

Drums and Brass

PMsD		
RF	1	1
RF	2	2
RF	3	3
RF	4	4
RF	5	5
RF	6	6
RF	7	7
RF	8	8
RF	9	9
RF	10	10
RF	11	11
RF	12	12
RF	13	13
RF	14	14
RF	15	15
RF	16	16
RF	17	17
RF	18	18
RF	19	19
RF	20	20
RF	21	21
RF	22	22
RF	23	23
RF	24	24
Reed 1 Lo	U89	25
Reed 1 H	KM184	26
Sax 1	E608	27
Reed 2 Lo	U89	28
Reed 2 H	KM184	29
Sax 2	E608	30
Reed 3 Lo	U89	31
Reed 3 H	KM184	32
Sax 3	E608	33
Bass	Notion	34
Violin	Notion	35
Strings	Notion	36
Horns	Notion	37
Harp	Notion	38
Foots 1/2	2x (Y) 4060-Boundary	39
Foots 3/4	2x (Y) 4060-Boundary	40
Keys L		41
Keys R		42
Band Verb		43
Special Verb		44
Conductor 565SD		45
VOD 565SD/EVO 500		46
VOD 565SD/EVO 500		47
SM Announce 565SD		48
Lead Verb		FX1L
Chorus Verb		FX2L
SFX 1		FX3L
SFX 2		FX3R
SFX 3		FX4L
SFX 4		FX4R
Drums L	DM1K	St1L
Drums R	DM1K	St1R
Trumpets	DM1K	St2L
Trombone	DM1K	St2R
Proj L		St3L
Proj R		St3R
COL/Zune L		St4L
COL/Zune R		St4R

FOH

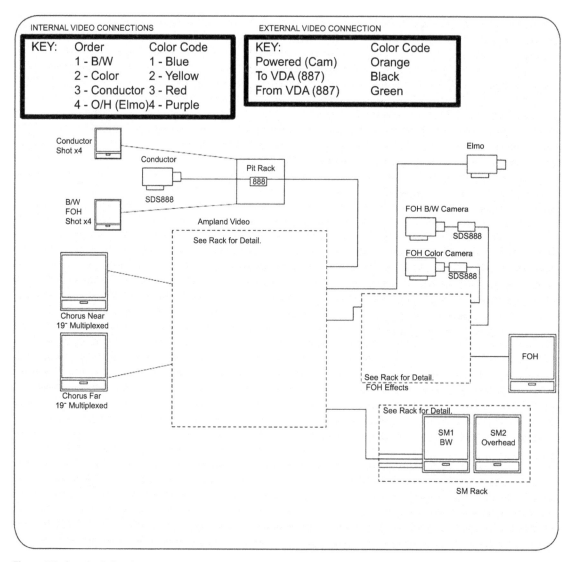

Figure 5.3 A typical signal flow intercom drawing from a sound designer.

include more drawings, but that is a matter of preference with the designer. These basic drawings are the foundation needed for a system to be built. Everything else can be fleshed out by the production sound or the mixer.

The other important task the design team has is to create an equipment order and bid it out to the sound shops. All Broadway shows bring their own system into the theater. On very rare occasions the producers will buy the sound system.

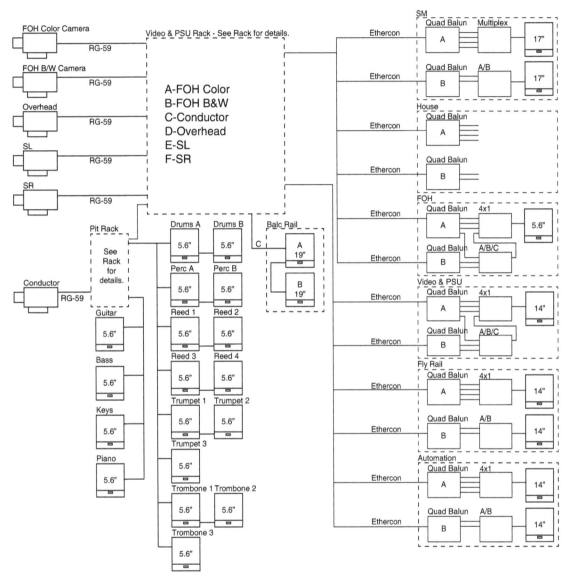

Figure 5.4 A typical signal flow video drawing from a sound designer.

Hairspray, The Producers, and *The Lion King* are shows that own their sound system, but this is definitely not the norm. More often the equipment is rented from a sound shop. Currently, the major sound shops in New York for Broadway are PRG, Sound Associates, and Masque. The designer is responsible for creating an equipment order that can be bid on by these shops. The shop with the lowest bid will get the show, as expected. The list of equipment the designer puts together is not a complete list of every item needed for the system, but rather a list of the big-ticket items. It includes all of the equipment in the signal flow drawings.

The list may or may not include the cable and mults for the system. Shops are very used to this and bid based on the major components of the system. They are used to estimating the cable and other parts and pieces needed to finish off the system, but as the shop time gets closer, the shop will want the list fleshed out to include every detail about the sound system. It is very important to get the big-ticket items on the list from the beginning. A common saying is that it is easier to cut than add. Typically, what happens in the bidding process is that the shops go as low as they can to win the show. Once they get the bid, they are not expecting major changes to the order. If you forgot to put a couple of Meyer CQ-1s on the order and you add them after the bid has been awarded, then don't be surprised when you get a call from the production manager asking why the shop is calling and trying to raise the rental cost based on changes to the system.

Accuracy counts and it costs. You won't make any friends if you get to the shop and double the budget. At the same time, if it was on your list and the shop agreed to it, then they are required to supply you with that gear. I am currently building a show and the designer included a specialty speaker with a note that said "with proprietary cable" as a note about the speakers. Once we were in the shop, the shop did not want to supply the cable. Instead, they wanted to make it a perishable, which meant that the production would have to pay $300 for a spool of cable. The designer simply called the shop and pointed out that it was noted on the order and they agreed to his list. The shop acquiesced and gave us the cable as part of the rental.

I designed a tour of *A Chorus Line* and the production manager asked me to put a Genie lift on my order because he was having trouble getting it from the lighting shop. Sound shops typically do not have Genies, but I put it on my order. After a shop won the bid, they realized that there was a Genie on the order that they had overlooked. They weren't happy about having to supply it, but they did, and acknowledged that it was their mistake for not noticing it. In the end it was worked out so that no one was upset, which is the objective. It is rarely worth a fight.

You have to be careful with arguing about your list and what you are entitled to. If you are going to insist that the shop supply you with something on your list, and you throw a fit that they agreed to it and have to supply it, then be prepared to have a hard time getting the amp you forgot to add to the list or the speaker the director asked for after the bid was awarded. If you are going to put up a fight for something, you better make sure you haven't missed anything. Sometimes you have to barter and trade. Keep in mind that the shop is trying to make money as well as make you happy. It is a hard balance. On one project I was getting close to the end of the build and had a stack of cable and gear worth about $20,000 that I didn't need, and I found out I was short a $100 SM58. I asked the shop for the mic as an add, and I was told the budget was so tight that they couldn't supply me the mic. I responded by stating that I had intended to give them back all of this gear that was on my order that was no longer needed, but since I couldn't have the mic, I would keep it instead. The shop gave in and I got the microphone.

Figure 5.5 shows a page from a typical equipment order. This is a detailed list of the gear needed and the quantity. The shop is not the enemy when it comes to the list. The shops are good at looking these lists and making guesses about what else might be needed for the system and what it will cost. If the shop does their job, they have anticipated the thing you might have forgotten from your list and will not be surprised unless your adds become out of control and expensive. Just keep in mind that it is a business and the shop has to make money. The worst equipment order I have ever seen listed only the big-ticket items like amps and speakers, and then had a note that said, "And everything else to make a functioning sound system." Not the best way to do an equipment order, as it leaves you without a leg to stand on when you try to point out the necessity of some expensive piece of equipment and the shop says they don't agree that it is required to build a functioning sound system.

CONSOLES

Qty:	Make:	Model:	Description:
1		EZ-TILT	
1	AVIOM	16/O-Y1	AVIOM CARD FOR YAMAHA DESKS.
3	AVIOM	A-16D PRO	REUSE
16	AVIOM	A16II	AVIOM A-16II PERSONAL MIXER
9	AVIOM	EB-1	EXTENSION BRACKET
2	YAMAHA	DCU5D	
2	YAMAHA	DSP5D	
1	YAMAHA	PM5D-RH V2	
2	YAMAHA	PA CABLE PM5D	CONSOLE POWER SUPPLY CABLE YAMAHA PM5D

HAREWARE/YOKES

Qty:	Make:	Model:	Description:
2		19" MONITOR YOKE	
8		FLAT STEEL FOR FRONT FILLS	PIECE OF STEEL USED TO MOUNT SPEAKERS TO THE DECK.
3		M1-D BUMPER	
9	AVIOM	MT-1	MOUNTING BRACKET.
14	MEYER SOUND	MM-4 YOKE	
14	MEYER SOUND	UPM YOKE	YOKE FOR MEYER UPMS

HEADPHONES

Qty:	Make:	Model:	Description:
5	SENNHEISER	HD480	SENNHEISER HD480 OPEN-AIRE HEADPHONE
11	SHURE	E5	SHUR3 E5 MONITOR SYSTEM EARPHONE
14	SONY	MDR-7506	HEADPHONE. WITH THAT STUPID SCREW-ON 1/4" ADAPTOR.

INTERCOM

Qty:	Make:	Model:	Description:
4	CLEAR-COM	AMS1025	CLEAR-COM AMS1025 AMP STEREO MONITOR SPKR
4	CLEAR-COM	CC DC BLOCKER	
1	CLEAR-COM	FL-1	CLEAR-COM FLASH BOX
9	CLEAR-COM	HS-6	TELEPHONE STYLE HANDSET WITH HANGING CLIP.
2	CLEAR-COM	KB-211	
1	CLEAR-COM	PT-4	HANDHELD CB-STYLE MICROPHONE.
5	CLEAR-COM	RM-220	TWO-CHANNEL REMOTE INTERCOM STATION
10	CLEAR-COM	RM-440	FOUR-CHANNEL REMOTE INTERCOM STATION
4	CLEAR-COM	RS-501	SINGLE-CHANNEL BELTPACK
5	CLEAR-COM	RS-502	2-CHANNEL BELTPACK
4	CLEAR-COM	SB-440	FOUR-CHANNEL ASSIGNMENT MATRIX MAIN STATION
1	CLEAR-COM	TW-12B	CLEAR-COM TW-12B CLEAR-COM/RTS SYSTEM INTERFAC
1	CLEAR-COM	TW-40	CLEAR-COM TW-40 2 WAY RADIO INTERFACE
2	CLEAR-COM	YC-36	CUSTOMIZE 2-3PXLR TO 7P MULT RS502
15	SENNHEISER	HMD-410	DOUBLE MUFF HEADSET

MATCHING DEVICES

Qty:	Make:	Model:	Description:
16	MISC	PHASE	XLR IN-LINE PHASE REVERSER

Figure 5.5 A typical equipment order from a sound designer.

In the end, though, the shop will try hard to please the designer. Designers develop loyalty to shops and when the designer feels taken care of he will want to go back to that shop. The bigger the designer, the more the shop will try to please the designer, because they know the designer will push for their shop on the next show. There are times when the shop is less concerned about the profit on a single show and more concerned with the impression they make on the designer and the production manager or technical director. If this is the case, the shop will give you just about anything you want because they are looking at a bigger picture. This always makes for a pleasant build. If this stage is dealt with properly, the design team will establish a good relation-ship with the shop and provide them with an accurate list that allows the mixer to build the show without too many conflicts.

6

MIXER'S JOB

The job of the mixer is always to fill the holes with whatever hasn't been done and do it efficiently and without need for praise. The mixer is there to make sure the designer's vision comes to life and to make everything feel effortless and smooth. There are mixers who pride themselves on how little the designer has to do. In the end, the less the designer has to worry about the little details, the more he or she can concentrate on the big picture. If the designer is trying to help fix intercom problems, he can't focus on why that reverb isn't working. When a mixer does his job right, the designer is relaxed and only has to think about the sound of the show. The mixer's job is simply to make the system work without drawing attention to the sound department.

If a mixer is lucky, he will be handed a stack of drawings when he starts on a project. This is not always the case, but for most Broadway shows this is the standard. The first thing the mixer needs to do is evaluate where the design team is and how he can help in the process. It is important to understand the team you are working with and what their strengths and weaknesses are. It could be possible to work with a brilliant designer who is just not good at system design and will need lots of help, but it is also possible to work with a well-oiled machine of a design team that cranks out show after show and has a system that requires very little input from the mixer. There are times when the design team will do a little and there are times when the design team will do everything. The rack drawings and cable order as well as the bundles and labels will sometimes be done by the design team. On larger shows with a well-established designer, this is definitely the case, but there are times when a lot of this will be left up to the mixer.

The first things that must be done to move beyond the broad strokes of the signal flow drawings are the rack drawings. All of the gear must be placed into racks for the show. As a mixer, it is important to know about racks. There are different size racks and they are sized by spaces. A 16u rack is very common, which is a 16-space rack. Racks come with rack rails that allow equipment to be screwed into the racks.

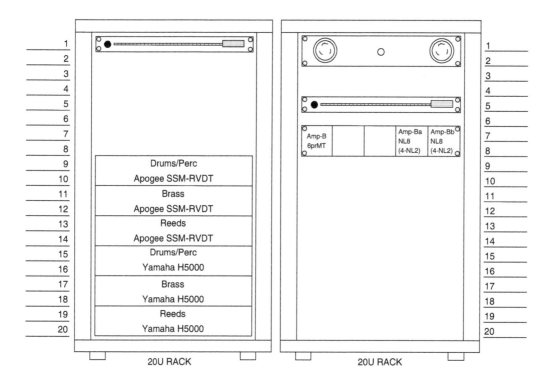

1. Do not use blanks or fans unless noted.
2. Do not zip tie power cords together unless cords can be disconnected from the unit.
3. Label front and back of all units.
4. If the rack is drawn without castors it needs to be able to come out of the tray.
5. No substitutions without the designer's approval.

Figure 6.1 A typical amp rack drawing.

Most racks are 19" wide and 19" deep. There are deeper racks available for use when building some amp racks. All of the equipment for the system will need to be racked up and the drawings show how it is to be racked. Figures 6.1 and 6.2 show some examples of rack drawings.

A rack drawing is required for every rack in the show. Rack drawings show where the gear for the show will live and detail the plan for the sound system so that people can build the system. Rack drawings usually consist of a drawing of the front and back of the rack as well as rack space

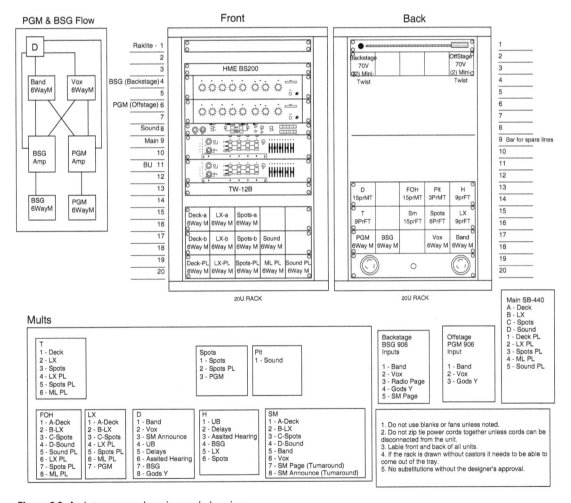

Figure 6.2 An intercom and paging rack drawing.

numbers. Sometimes the drawings will detail patching and wiring needs for the rack, as well as the names for the equipment and the equipment channels. The more detailed these drawings are, the better. The more information you provide your build crew, the less they will need to ask you. The fewer questions they have means more time they can work on their own while you deal with problems they haven't gotten to yet.

As the rack drawings are done, the next level of the system starts to become clear, which is the way the system is connected and the parts and pieces needed for the system. The supplemental equipment order for the show starts to grow as the drawings are done, which consists of the racks you need as well as the cables and mults and other minutia needed for the system. The important thing to know is that the shop will rarely bring you anything you didn't ask for. The shops are very good at keeping track of every little item you take out of the shop. The sound shops are large companies dealing with tens of thousands of items, and every item has a value. If you need something, you will have to ask for it. Never assume you will get anything. An SM58 will always come with a mic clip, but a Neumann U87 will not automatically come with the $400 shock-mount clip. It is important to do the drawings and plan out every detail of the sound system so you can give the shop a very accurate supplemental equipment order to complement what the designer has asked for.

The shops are very accustomed to getting a list from the designer that is the broad strokes of the system and includes the big-ticket items, and then later getting a list of the nuts and bolts of the system. The shop will let you know if you have overstepped the budget if you ask them to. They understand that the mixer does not have the authority to add anything that will increase the rental price of the equipment and they will give you whatever you need as long as it does not cross that line. If it does, you will have to find a different way of working out the system or you will have to explain why this is so crucial.

7

THE SHOP

Once all of this has been worked out, the next step is for the mixer to go to the shop and build the show. This is where the system starts to become the mixer's, and this is where a mixer starts to prove his worth. In the shop, we label and rack all the equipment and cable. We also plug the entire system together and test it. Most mixers are very particular about how racks are built and what they look like. Mixers can also be very picky about how racks are labeled and patched and how the spare cables are tied up. Good mixers pride themselves on their finished systems. The look of the system is a representation of the mixer, and the cleaner and more thought out it is, the better a mixer looks. Nothing is worse than having another Broadway mixer stop by and look at your system and have it look a mess. It pays to take pride in the way your system looks. As a mixer, you are showing people your skills and your work ethic when they see your system. If your mix position is a mess, then they will think your mix will be a mess as well, and people will be less likely to want to work with you.

To be successful in the shop, you have to understand the world of the shop. Just as you must understand the hierarchy of the design team, you must also understand the hierarchy of the shop. You have to understand who you can talk to and how to ask for something. You have to understand the timing of the shop and when to get concerned. You have to understand the order in which gear is going to come to you and how to deal with it in that order. You have to understand how to orchestrate the build. We will go into great detail about what the crew actually does in the shop in Chapter 9.

Most designers have no interest in the shop or the process of building the show. It is not that they do not care; it is not their job. A majority of Broadway musical designers were at one time mixers and are probably more than capable of prepping a system in the shop, but when they are in the designer role, their place is not in the shop. That is not to say that the designer is not important to the shop build. In fact, because the designer is removed from the build, he can actually be a very strong asset. Most issues that arise in the shop can be dealt with by the mixer, but some require the clout of a designer to accomplish. When those issues arise, it is time to call the designer. The designer can be a major asset in the shop because he can be the heavy. The designer can be the one who calls and pitches a fit so you don't have to. The result is that you get what you need and no one has an issue with you. If you throw the fit, then the shop could turn on you and you could find it hard to get anything done.

Just as in any field, there is a hierarchy in the shop. It is important to understand the key positions and how they relate to the build. It is important to know who to talk to when you have a problem. It is also important to know who to talk to when you need some equipment. Shops have a very structured system of building shows. If you don't understand the shop's system, you are bound to get in trouble. You need to understand what a build zone is and what the foreman does and what the salesperson is responsible for. It could take a dozen builds to fully grasp how shops work. In this section, we break it down.

Salesperson

The first person you are likely to talk to at a sound shop is a salesperson. What's odd is that this person is called a salesperson but doesn't really sell anything. The title is a little misleading, because the salesperson is in charge of making sure all the equipment needed for a show is available. This is the person who will take the designer's equipment list, put it into the shop's computer system, and book equipment for your show. A good salesperson is an incredibly important part of having a successful build. A bad salesperson can leave a build in shambles.

The salesperson is in charge of getting approval to order equipment for the show and for ordering the equipment. If the salesperson does not order equipment on time, the show could be in real jeopardy. On one show I did we were using a Cadac, which is still the best-sounding and best-feeling console you can mix on. A Cadac is an analog console with some automation built in. To make the automation function, you need a computer running a program called Séance and a piece of equipment called a CCM (Figure 7.1). The CCM is the Central Control Module for the console. Without the CCM, the console will not connect to the computer and there is no automation. You are left with an analog desk. The salesperson for the show forgot to order a CCM and by the time I found out, the delivery time for a CCM was scheduled for our first preview. It was a disaster of a situation and led to many phone calls. Luckily a solution was found, but it shows how the neglect of a salesperson almost took down a first national tour.

Every sound shop has some form of database software that tracks their equipment. This software keeps track of every piece of gear owned by the shop and when it is booked and when it is available. In the case of larger shops that have offices all over the country, this software also has to keep track of equipment in other parts of the country. The salesperson is responsible for booking equipment for your show. The salesperson takes the designer's list and any other supplemental lists and inputs it into the shop's system. As the salesperson does this, he or she can see how many items will be available for your show. The person books the equipment so it will be locked out for any other show.

This is an easy process if the shop has all of the equipment on your list in stock. It becomes increasingly more difficult as they find gear that is not available. They then have to look to see if it is in another shop and if it is more cost-effective to ship it or purchase a new one. Then they start looking at

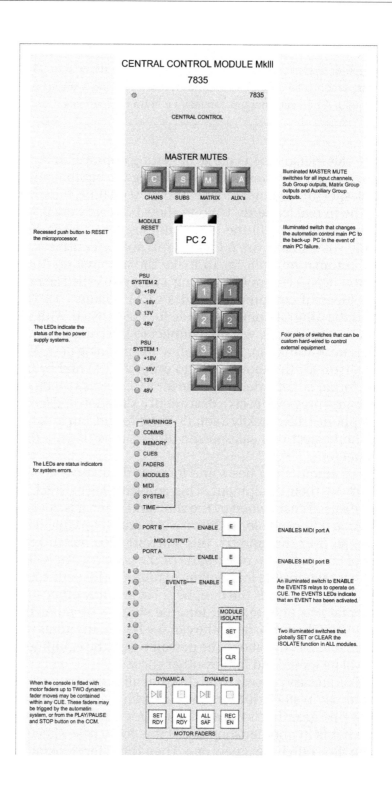

Figure 7.1 A Cadac CCM.

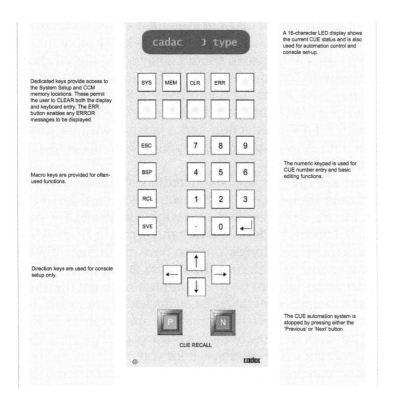

Dedicated keys provide access to the System Setup and CCM memory locations. These permit the user to CLEAR both the display and keyboard entry. The ERR button enables any ERROR messages to be displayed.

Macro keys are provided for often-used functions.

Direction keys are used for console setup only.

A 16-character LED display shows the current CUE status and is also used for automation control and console set-up.

The numeric keypad is used for CUE number entry and basic editing functions.

The CUE automation system is stopped by pressing either the 'Previous' or 'Next' button.

Figure 7.1 Continued

your show dates. It is possible they will have the equipment but not until you leave the shop. When this happens, they will ask you if they can drop ship the equipment to you and most of the time this is acceptable. The next step is for them to buy the equipment needed. This requires approval by people above them, which usually requires the salesperson to put together a list of all purchases to analyze whether the show is costing too much. At this point it is common to get a call asking about substitution and most of this is done in the bidding process.

Once the equipment order has been fully booked in the shop's system, it is up to the salesperson to keep tabs on the status of equipment deliveries for the show. That person is also expected to look out for missing holes in the equipment order. It is not uncommon for a salesperson to point out pieces of equipment you forgot to put on your list. A salesperson is not required to be a mixer or a stagehand. A salesperson may have never built a show or mixed before.

None of that is part of his or her required skill set. The salesperson's job is to know gear. Salespeople know more model numbers than any human should. They could literally hold a conversation for minutes using only numbers and letters. They may have never used an Aphex 1788a and they may have no clue how to use it, but they know that there is a version with a LAN connection and a version with an RS422 connection. They also know that you have to use a specific router with the units.

Good salespeople are an incredibly valuable resource and should be treated with respect. It is a hard lesson to learn, but the salespeople are on your side. Really, the only things not on your side are money and time. Given an abundance of those two things, you can have anything you want for your sound system, but when those are in short supply, then decisions have to be made about how to get the show up and running. Sometimes those decisions are tough, but it is important to know that the salespeople are just messengers. They cannot allow or deny a piece of gear for you. They can merely book it or purchase it for you.

At a certain point, the salesperson will start talking to the mixer to find out about the supplemental equipment needed for the system. It is at this point that the mixer takes over the list and must maintain a good record of what is needed and when it is needed, and must communicate changes that arise as well as mixing items from the list as soon as possible. The first task to be done in the shop is to have the shop print out a list of what is on your order and to cross-reference the shop list with the designer's list. It is important to know from the start of a build that the shop is fully aware of every item needed. The shop runs like a machine, and people all over the shop are pulling equipment to fill your order. If the list is accurate, then you can count on getting most, if not all, of the items on your list. As the shop pulls the gear, they mark it in their system so they can constantly see how many items you ordered and what you have been given and what you are still owed. If the two lists are accurate, this process goes smoothly. If you do not take the time to check the shop list, you could find out halfway through your build that a clerical error was made and only half the speakers are on the shop list, which could put you in a bad place if they do not have the speakers in stock. Making sure these lists match is a tedious process, but it is crucial to a successful build.

Once the rental list has been hashed out, the salesperson will want to go over the perishables list, which is a list of equipment the show will need to purchase. This list can include all wireless mic supplies such as batteries and canned air. It can also include items that were on the designer's rental list but that the shop considers perishable. The perishables list can also include items such as P-Touch tape and electric tape. There are times when it is more cost-effective to buy the perishables somewhere other than the shop. The shop usually does not care if you buy the perishables somewhere else, and the money you save can get you some brownie points with the production manager.

Shop Foreman

Most of the sound shops around New York City are union sound shops and have a similar structure. The sales department is the non-union front office part of the shop, while the show build area is the union part of the shop. This area usually looks like a giant warehouse, and the person in charge is the foreman. The shop foreman is a key player in how well the shop runs. The shop foreman meets with the sales department and goes through the paperwork supplied by the design team. It is always a good idea to pass on drawings of the sound system to the shop as early as possible so the shop can understand the scope of the show.

The shop foreman looks over the scope of the show and lays out a plan for the build. Most shops have specific show prep areas, and the shop foreman books shows in the build areas, or zones. The shop foreman also assigns a key to the show and communicates his concerns to the sales department so that those concerns can be discussed with the design team. The shop foreman also watches out for the overall safety of the people in the shop as well as the people who have come in to build and the equipment. It is not uncommon for the shop to tell you that you have to change the way a rack is being done because it will be better for the gear another way, and since the gear is a rental and belongs to the shop, you should listen to their requests. If you don't and the gear comes back to the shop broken, your show is going to have to pay for the equipment. The shop foreman also makes sure trucking has been arranged.

A well-run shop is the work of a good foreman and a pleasure to work in. When the foreman is less than stellar, it can really hamper a show. Gear might not show up to the zone on time because the foreman didn't book the right labor to pull your show, or the gear could be lost because the foreman hasn't had the shop people cleaning the shop enough once shows return. When you are in a shop and you are digging through racks of untested gear that just came back from another show just to get your system built, then you know there is a problem with the foreman. Odds are the next time you come back to the shop it will not be like that, because no shop can survive for long in a state like that. Luckily, the shops employ some top-notch foremen who are knowledgeable and helpful.

Show Key

When you arrive at a shop to build a show, you will be introduced to your show key, who will take you to your build zone. The show key, sometimes called the show captain, is the most important person in the building to your build. The show key's job is to bring you equipment. The shops do not want you wandering through the shop picking up gear on your own. They want to bring it to you so they have a count of what you have been given. The show key takes the shop equipment order and pulls it and brings it to your zone. The show key usually starts pulling your show several days before you arrive so you will have gear in your zone waiting for you. No one likes walking into an empty zone.

The show key is your conduit to communicating with everyone. If you need to talk to the foreman, you tell the key and he will bring the foreman over. The show key does not have the authority to add equipment to your order, so if you need to add something the show key will bring the salesperson over. If you need to know when the console will come into the zone, the key will check with the console department to find out. All shops are divided up into different departments. There is an amps and speakers department, a video department, a microphone department, and several other departments. There are people in these departments who are all looking at your order and pulling, prepping, and testing equipment for your show. It is the key's job to communicate with those departments and make sure everything is on track. A good key can head off problems without you even knowing about them. He can look over the drawings and paperwork for the show and anticipate problem areas. A good key can also point out things that may have been overlooked, forgotten, or planned incorrectly.

For this machine to work the way it should requires diligent planning and paperwork for the design team. An accurate equipment list is crucial. Without an accurate list of what the show needs, the build will be hamstrung. Time is also crucial. The most successful builds happen when an accurate list and accurate show drawings are supplied to the shop weeks, if not months, before the prep begins. This gives the shop time to communicate with the designer and come up with adequate substitutions, and then time for the salesperson to enter and book all of the equipment. And then time for the foreman to develop a plan for the build. And then time for the show key to look over the paperwork and start pulling the equipment. When done right, the mixer can even suggest what he would like to see in the zone on day one.

One thing that can mess this system up is the bidding process. There are times when the bidding process drags on and runs right up to the day of the build. There have been shows that didn't know what shop they were going to until the day before the build. When this happens, everyone is bound to be frustrated and unhappy with the build. If you walk into a build zone with nothing in it, you should take the time to understand why. Was it that sufficient time was not given to the shop to allow them to prepare for the build? If so, then you have to find a way to be patient. Was it that adequate paperwork was not given to the shop, so the shop is not fully aware of the equipment needed for the show or the scope of the show? If so, then you have to help get the shop the information they need, and you have to understand that it is not the shop's fault. Was it that the shop underbid the other shops to get the show, but they are too busy to deal with the show and just built a dozen shows and have no gear? If so, then it is going to be an ugly build with the shop defensive and everyone else angry.

No matter what the reason or the state of the shop, the mixer's job is to build a sound system. An angry mixer does not accomplish much. A mixer has to make the best of whatever situation he or she has been dealt. You can complain all you want about the shop, but the fact is that you were hired to make it work. The designer does not want to hear excuses. The audience and the director are not going to hear that you had a bad time at the shop. Your job is to do your job and make it look easy. That will get you hired again.

Shop Etiquette

There are five lessons I have learned about dealing with sound shops. I have spent a great deal of time in sound shops in every position possible. I have been a designer and an assistant. I have been a mixer and a production sound. I have also been label monkey number three and box pusher. Working in rental shops is an integral part of doing sound and it takes some getting used to. The different positions require different tools and sensibilities. Dealing with shops is a skill that is learned through mistakes, and every mistake takes time to be forgotten. Without the shop, you have no gear. Without the shop's support, you have no help. Without the shop on your side, it is a steep hill to climb, but once you get comfortable working with the shop, things get much easier. So after many personal mistakes, here are the five lessons I have learned about sound rental shops.

Your Lack of Preparation

This is a classic stagehand adage. Your lack of preparation does not make this my crisis. Sometimes expressed as "Not my problem," at one theater where I worked it was just simply "NMP," which was printed in giant block letters near the pin rail. And it is as true as can be. When I was just getting started, I ended up in the shop on several occasions with three days to do 15 days of work, having been hired the day before. I made the mistake of going to the shop in crisis mode. I would be wound tight and agitated at everything. "Why don't I have this? Why isn't that done already?" It took me time to understand that the people working in the shop work there every day, and this problem was mine and not theirs.

The Shop Is Not the Enemy

My first experiences in the shop were as just show labor. I worked under other people building shows and learned some great lessons from them. I learned how to build a show and where to put labels. Unfortunately, I also learned one lesson that was not so good and took me quite a while to unlearn. That lesson was the shop is the enemy. The truth is the shop is *not* the enemy and you want them on your side. It is easy to get into arguments with the shop and become unreasonable and demanding. Man, I made that mistake too many times, and it gains you nothing. It is such a cautious balance trying to get your show built on-time and under-budget with so much out of your control. Just remember, you will get the Cadac and the RF rack just in time to push it on the truck, but you will get it.

Learn Your Terminology

Steck Rails. G-Blocks. Mults. Bundles. Waber strips. Show Key. And much, much more. There is so much terminology involved with building a show, and it takes time to learn. Different shops have different terms and different parts of the country have different conventions. In New York shops, a bundle is a group of cables taped together. In other places it is called a loom. Power strips are called Waber strips because that was the manufacturer at one time and the name stuck. Same thing with a Sammy, which is a big wooden box affectionately named after the maker at some shops. This minutia is important to know. It makes working with the shop so much easier if you can talk to them in their language.

Build a Relationship

When I got my first big design job, I thought I was the stuff. I thought, "Look at me. All the shops are going to be clamoring to get my show." Woof, was I wrong. I put together my equipment list and sent it out to all the shops and then waited by the phone like a schoolgirl after a first date. And then nothing. No one called. So, after a few days, I called them and found out that no one was interested in my little show. I then started panicking. I had a tour to design and I needed equipment and I had nothing. Finally, I found a shop that was willing to do the show, and that was when I realized how important it is to build a relationship with the shops. Part of the reason is that until you have established yourself with the shop, they have no idea how legit the show is. They don't want to waste their time if the gig isn't going to happen. Also, even though I thought I had a big fat show, looking back now I realize that the budget was horribly low, so it was no wonder no one wanted it. Now that I have done lots of work with the shops, it is easier. It is also getting logistically easier. The more I work with a shop, the more they know what I want and my shortcomings, which makes for a simpler process. At this point I prefer going to a shop I have worked with because it is much easier than starting from scratch.

Have Fun

I have found that the more light-hearted I am, the better things turn out. If I roll with the punches, I seem to get punched less. If I am open and will allow substitutions, then things are smoother and I usually get the system built faster. If I can accept that other people might have a better idea than I do, then I usually end up with a better product. I have good friends at the shops and look forward to going and building shows. I know there are some people who remember some of my youthful mistakes and haven't forgiven me for them, and I don't blame them, but you don't make omelets without . . . well, you know. I am still learning the rules and trying to get better at the process of building shows and working with the shops. Maybe one day I will perfect it, but until then I will try to laugh at my mistakes. After all, it is only theater.

8

BUILD SCHEDULE

So, it is time for the shop build. Hopefully you have done all of the prep work to make the build go smoothly. Hopefully you have more answers than questions. Hopefully you are ready. A show build time depends on the budget and size of the show. A typical Broadway show builds in three weeks, although sometimes you can get four. Some shows are smaller and have a smaller budget and may only give you two weeks to build. Some large shows, like *Spiderman*, build for months. These are rare. Whatever the length of your build, the one thing you can be sure of is that it will not be enough time. Somehow the powers that be can judge a show and they can somehow schedule the build just short of what you really need, which ends up causing you to push and get it done just in time.

Production Manager or Technical Director

Every show on Broadway has either a production manager or a technical director. These positions are basically the same and only differ in title. The technical director is usually referred to as the *tech*. This is the person who was hired to take care of all of the technical aspects of getting a show prepped, built, loaded-in, and open. Basically, everyone answers to the tech. He is the master of the budget and the schedule. He hires most of the crew. He will coordinate truck deliveries and crew numbers. He will have the final say on big decisions. A Broadway tech is very different from the technical director in college or at a regional theater. The tech is really the boss of everything and is not involved in the actual construction or physical installation of the show. He is in charge of the calendar and the meetings for the production. He is similar to the production manager at the college level, except with the added responsibility of being the technical expert as well.

The tech does not work on the production on a daily basis after the show opens. He is rarely seen after the opening night, but he is still the go-to person for technical questions. If the show needs to have a big and expensive workcall to fix something, it will need to be approved by the tech. If an expensive item needs to be purchased for the show, it must be approved by the tech. Even though he may not be in the building, the tech is still in charge of the production.

The tech is the person who hires some of the running crew for the show. The tech usually hires the sound mixer, but the tech does not really get to make the decision about who the mixer will be. All mixers are picked by the sound designer, but the sound designer does not have the authority to officially hire or fire a mixer. Instead, the sound designer tells the production that he has picked a mixer and the tech will contact the mixer to work out a deal. Unfortunately, the mixer then has to negotiate a rate with the tech. I say unfortunately only because it would be nice if there were a set rate for a mixer on Broadway and it wasn't a negotiation. On the rare occasion when a mixer needs to be fired, it is a decision that must be made with the approval of the sound designer.

Hire Crew

The first task for the build is to hire a crew. Typically, it takes at least four people to build a show. There are times when the mixer is in complete control of who will work on the build, but that is not terribly common. More than likely, the designer will want certain people on the build. The house electrician might also want certain people on the build. After they have staffed the build, the mixer is free to fill the open spots with anyone he likes. Over time, you will build shows with people and will build friendships as well as common ground. Nothing is better than building a show with a crew that you have built shows with repeatedly. They understand the way you like to do things and the strange idiosyncrasies you might have. There are certain people you will just enjoy working with, no matter what. Picking a good build crew can make even the most arduous shop time more relaxing.

There are people who have made a career out of building shows. They bounce from one build to another and are top-notch professionals and you will get high quality work from them. These people will also cost and will want the going rate. Shop work is typically paid as a day rate and the top-notch people will want the prevalent day rate. As a mixer, you will want to know what the budget is for the build. Usually this budget does not include the pay for the A1, A2, and assistant. You will want to know how many people you can hire and how much they will be getting paid. Sometimes you will be given a rate that is low and may need to find a way to convert that rate into an overall budget and readjust the day rate.

For example, you may be told that the budget is four people for ten days at $100 per day. Let's assume the going day rate is $200 per day. Obviously, your budget is really low if you want to staff your build with qualified people. In this case, it may be necessary to talk to the tech about the budget. If there is no more money to offer, then you may need to suggest that you bring in fewer people at a higher rate. As long as you stay under the budget, this is usually not a problem. So in this case your budget is $4,000, which will break down to 20 slots you can fill.

If you look at the flow of the time in the shop, you will know that the first day in the shop is slow and there isn't much to do. The first day consists of setting up the build zone and checking the shop order with the designer's order and printing. It is usually not worth bringing lots of people in for the first day. If you bring no one in, then you will have the A1, A2, and assistant, which is probably enough. The last day in the shop is dedicated to packing everything up and putting it on a truck for delivery. This is another day that may not require a full build crew. It is also a day that does not require a highly specialized crew. It is easy work that cheaper labor can typically do. If you take this into consideration, then you can staff your build with three people for eight days at the higher rate and finish the build with two people.

Once you have figured out what you can afford and what spots have not been filled by others, it is time to start making calls. If you have been doing this for a while, you will have no problem filling the call. You will have more numbers and names than you will know what to do with. But if you are new to this, you will need to search people out. Take your time and find experienced Broadway show people. Try to find people with experience with the theater you are going to. The more your crew knows, the more they can keep you from making big mistakes.

Break Down System into Tasks

At first glance, a sound system seems straightforward. It is just some mics and speakers. But on further investigation, you realize it is incredibly complex. You may need a crossover rack. Your designer may require a distro rack or remoted pre-amps. Your designer may have power theories or multicabling ideas you have to abide by. Not only do you have to build a system that fits into the space available at the theater, but also a system that caters to the designer.

Since many musical theater designers are mixers themselves, they are going to be opinionated about the system you build. They may not like the way you built your racks or the way you laid out your gear. They may request certain criteria for your build, such as limiting you to only using certain size multicables. There is nothing better than working with a designer who used to be a mixer. They understand the rhythm and chaos of a show. For the most part, the higher up they get as a designer, the more they let go of the mixer, but no matter how far removed they get, they still have strong beliefs on how things should be done. You would be wise to learn what your designer likes and try to accommodate those ideas. In the end, those designers are still mixers, and many of them can mix circles around you. They are also still quite adept at troubleshooting.

I have fond memories of working on *Wicked*. I remember times when something would go wrong and the mixer, Douglas Graves, would start troubleshooting. Douglas is an excellent mixer with top-notch skills, but everyone gets stumped at some point looking for that one little button that wasn't pressed. Tony Meola, the sound designer, would occasionally say from his tech table, "I know what is wrong. If you want my help, just let me know." It was just so interesting that he was sitting there with the answer and was allowing his mixer the space he needed to fix the problem on his own. Occasionally Tony would be asked to help, and inevitably he would walk over and hit one button and instantly fix the problem.

It is important to remember, though, that a sound system is not just the speakers and the mics. In fact, that is usually the easiest part of your system. The real complication comes when you get into the other bits and bobs. You have to deal with the inputs and outputs of the sound system, and the intercom, video, and paging. This alone could overwhelm you and become the most complicated part of the sound system, and if this part isn't done correctly, then all of your hard work on the actual sound system will go right down the drain. It is crucial to understand that the most important part of your system is the intercom.

I have said it before and I will say it again: I would rather my speakers sound tinny during tech and have my intercom working flawlessly. I would rather the speakers buzz. I would rather they were on fire. I would rather have an ear cold and have my speakers on fire. I would rather the console discharge 2 volts every time I touched it and have an ear cold and have buzzing speakers in flames and have a pencil stuck in my eye, if it meant that the intercom was perfect. The reason is very simple. I can always apologize for sound problems. I can always say, "The fire department is on its way and we will mix again once we can touch the console." But no one wants to hear any excuse about intercom. If it doesn't work, nothing can happen and you have made no friends. The lighting design team will be annoyed and the production manager will be right there to tell you how much money is being wasted because com doesn't work and no departments can work.

The intercom system is evil. It's voodoo. It's a jilted lover with a long memory. Treat your com system with care and respect. Read the manuals for every item you are using in the intercom system. Test it daily. Test it several times a day.

Inevitably, the intercom system will be perfect every time you test it. It will work flawlessly and quietly until the first time someone needs it, and then somehow the stars realign and the whole system starts buzzing and everyone gets annoyed. The best course of action is to always assume it is broken, and then you will never be disappointed. It is possible to have trouble-free intercom experiences, but it requires forethought and knowledge of the equipment and dedication to the goal. And always have spares! Cables, packs, and headset!

Once you have mastered your ins and outs and your intercom system, next up is the paging system. Broadway theaters are considered a "four-wall" theater, which means when the theater is rented, it comes with no equipment. A Broadway theater starts every production as an empty theater that doesn't even have a paging system. You have to bring everything you need, including a paging system. A paging system usually includes a 70v amp and 70v speakers. It also includes cable and paging mics for the paging system. A 70v speaker will need to be hung in every dressing room as well as in common places such as hallways and the Green Room. It is an Equity requirement to provide paging, which includes good mics and a program feed, for the cast. If you do not provide adequate paging, it could be reported to Actors' Equity.

The video system is the next part of the sound system that has to be dealt with. To be clear, the video system we are talking about is show-critical, closed-circuit video to aid in the communication needed for a safe production. This is not video or projection that is part of the artistic side of the production. As a sidebar, I would just mention that cue lights are also an integral part of show communications but are not something sound deals with. Cue lights are handled by lighting and light shops. Typically, there will be a black-and-white camera on the conductor. This camera will need to be distributed all over the place. It will need to be sent to the Green Room and the stage manager's office as well as the calling station and to balcony rail monitors. If there are vocal signing stations backstage, the conductor shot will need to be sent to those locations as well. There is usually also a black-and-white infrared camera and a color camera shooting the stage from the house. These cameras are usually on the balcony rail. Infrared emitters will need to be placed over the stage to illuminate the stage for the blackout moments.

The conductor monitors on the balcony rail will also need to go through a blackout generator. The blackout generator is controlled by lighting and allows lighting to build a cue that blacks out the stage as well as blacking out the monitors on the rail. If you do not use a blackout generator, then if a video monitor loses signal, it will still emit light from a black screen. The blackout generator eliminates this light leak. Lighting will typically send a circuit to sound that they control so they can build this blackout into their cues. This box can live in your video rack between the send to the balcony rail and the feed from the conductor camera.

It is also possible to have several other specialty cameras for the production. These cameras will be dictated by the needs of the stage manager, the fly rail, and the automation operator. It is not uncommon to have seven or eight cameras on a show. Most of these specialty shots will not need to go anywhere but to the calling desk, which is where the stage manager will call the show, the fly rail, and the automation desk. Sometimes these specialty shots will need to be remote control cameras that can move and zoom and focus. In this case, you will need to figure out who needs to control the cameras and run the controller to that location. There are also times when multiple conductor cameras are needed.

In the case of *Spring Awakening* on Broadway, the conductor and musical supervisor, Kim Grigsby, had to move between two playing positions. When she moved, a different camera needed to be used so she was always looking into the camera. To accomplish this, we had to run both cameras to the stage manager's calling desk and pass it through an A/B video switcher that then sent the signal down to the video distribution rack. When Kim moved, the stage manager would switch to the appropriate camera. The result was a very smooth transition and everyone in the show could always see the conductor looking into the camera.

The next system you will need to consider is your RF system, which includes your wireless microphones, in ear monitors and wireless com, and walkie talkies. There are times when you will even need to deal with wifi networks interfering with your wireless mics. Zaxcom is a brand of wireless popular with some television broadcasts and can have a negative side-effect. Zaxcom will squash certain wifi signals completely. I saw a tech come to an immediate grinding halt one day when I turned on a Zaxcom unit and shut down the Internet for the entire theater.

With your RF system, you will want to check your system for intermodulation problems. Intermodulation is the hobgoblin of wireless mics. Intermod creates specific sounds. The first sound is usually called "birdies." It sounds like a fluttering. The next is partial transmissions from other frequencies. The next is full-on static. So, what causes intermod and how do you avoid it? There are several types of intermod. One type is when you overpower your antennas. This causes frequencies to be overly boosted and leads to static on your frequencies. The other type is when two or three transmitters are close to each other and those frequencies combine to create a frequency that is similar to a frequency currently in use in your system. This usually causes "birdies" or partial transmissions from other frequencies being heard on the wrong frequency.

To understand intermod, you can think of dropping a rock in water. When the rock hits the water, rings emanate from the rock. Those rings are like frequencies. If you drop two rocks in water at the same time and close together, the rings from each rock collide with each other and create a new set of rings. That is like an intermodulated frequency. The closer the rocks are to each other, the stronger the second set of waves created, which is the same with wireless frequencies. The same holds true for three rocks or three frequencies colliding.

It is favorable, necessary, and possible to have an RF system with little or no intermod problems. If your system is so large that there is no way to exclude all of the intermod issues, then it is possible to manage the potential intermod problems. Intermodulation can easily be figured out using simple math. What we care about is intermod created with two or three transmitters. We also care about the severity of the intermod, which is described with the word "order." Two-transmitter third order is an intermodulated frequency of the highest power for two transmitters. If you think of the rock example, when the waves emanate from the rock there are several rings. The first is the most powerful and as such would be like the third order. The second ring is weaker and would be like the fifth order. We don't care about the orders past the fifth because they are usually too weak to interfere with our system. We also do not care about even orders such as second and fourth, because the math of even orders creates frequencies that cannot be in our system.

An example of the math is as follows:

Frequency 1 (F1) = 525.000
Frequency 2 (F2) = 530.000
Frequency 3 (F3) = 520.000
Frequency 4 (F4) = 535.000
2(F1) − F2 = 2 Transmitter 3rd Order Intermod
1050.000 − 530.000 = 520.000

This shows us that these two frequencies can create a third intermodulated or phantom frequency of 520.000. If F1 and F2 are close to each other and frequency 520.000 is being listened to, then you will probably hear some form of static.

An example of 3 Transmitter 3rd Order math is:

(F1) + F2 − F3 = 2 Transmitter 3rd Order Intermod
525.000 + 530.000 − 520.000 = 520.000

This shows how these three frequencies can create an intermodulated frequency that could be heard on frequency 535.000.

The math involved to find intermodulated frequencies is very simple, but the number of frequencies is staggering. If you have a system of 26 frequencies, there are almost 80,000 math calculations for the intermod of that system. To test for intermod, you need a computer program such as a program I wrote called RF Guru or a program called IAS (Intermodulation Analysis System Software). If you have a system with potential intermod problems, you can still manage the problems to make sure frequencies are not onstage at the same time that can cause the problems. If you do not check your system for intermod and you never have any problems, then you are just getting lucky. It is imperative to know your RF system and make sure it is solid. Most of the time the shops will do this work for you, but it is important for you to understand it and have some ability to test it yourself, because you just cannot predict everything you are going to encounter when you get into the theater.

I was a sub mixer for *Legally Blonde* and subbed for a great Broadway mixer, Bob Biasetti. I also subbed for Bob on *Dirty Rotten Scoundrels*. I had a problem one night while mixing *Legally Blonde* where the lead's mic let off a huge bit of static. I immediately thought it was intermod. I had the shop send an RF tech to the workcall the next morning to test the system and told him I was worried it was intermod. He said it didn't seem possible. He had done the math and the show had been open for almost a year, so he didn't think it could happen. But I had done the math as well, and I had found a potential conflict. I explained that I thought it was intermod because I was using the lead's backup mic, which we rarely use, and we were in a scene where the entire cast was clumped dead center and the lead ran right in the middle of the clump to sing and then loud static. He still didn't buy it. So I powered up the entire RF system and brought it out front. I soloed the backup mic and moved the packs I thought were the problem near all the other packs and, sure enough, we heard the same huge static sound. It was an intermod problem that had been there since day one, but we never heard it because we had never listened to that mic at that moment. And it was something that could be predicted using math. We switched that pack to a different frequency and the problem was solved.

Once you have made it through these systems, you can move on to cabling and bundling, which we get into in the next chapter.

Payroll

If there is one aspect of the build process that has to be addressed and taken very seriously, it is payroll. As the A1, you are the point person people are going to look to when it comes to payroll questions. You will be required to fill out timesheets and give them to someone, which is usually either the head electrician or the production manager. You will also be responsible for getting tax forms filled out for all the people on your build crew. You will also need to know the deadlines for getting this paperwork filed. If done properly, you will have no problems and people will work with you again. If not done correctly, it can really hurt people.

It is important to understand who works on a build crew. The people who build shows are all independent contractors, which means they are usually working on a 1099. A 1099 is a tax form that shows the amount made by a person from a company with no taxes withheld. At tax time you file your 1099s and have to pay the taxes on the money from those 1099s. As independent contractors, these people do not have the most consistent income stream. They are hired for a specific amount of time to build a show. If they can continue to get on a show build, then they are fine, but there are times when the build work dries up. This business is really hard on people. You are constantly chasing work; the common term in the business for going from one short-term job to another is "bounce." The result is a constant need to network and a good ability to save a cushion of money when you have work so you don't go broke when there is no work.

When you build a show, you are more than likely going to be hiring friends. The people on the build rarely know or meet the production manager or producer or anyone else involved with the show. Their only connection to the show is the A1, and they are dependent on the A1 for all communications with the show. If you make a mistake in your payroll and someone doesn't get paid correctly or on time, you could seriously jeopardize your friend's livelihood. This is of utmost importance to me. When I start a build, I find out what my budget is for crew and I figure out how many people I can hire. I then call the people I use regularly to fill the build. On day one I get the tax forms out of the way and I do payroll for the entire build and turn it in. I have learned over time that different companies have different policies on processing checks. There are times when I have to turn the paperwork in during week 1 of the build, it gets processed in week 2, and checks are mailed in week 3. If I wait until the second or third week, then people will not get paid for weeks after the build is over.

It is not uncommon for things to slip through the cracks. It is also not uncommon for a production company to be slow to pay. There are times when I will get calls weeks after a build asking where the check is and I have to call the person in charge to find out. It is the responsibility of the A1 to stay on top of this to make sure people get paid. I had one build that went horribly wrong with payroll, but fortunately not because of me. The build had been done for several weeks and we were deep into tech, and I had a friend who was on the build call me in a panic. He had not received his check and he could not pay his rent; he was bouncing checks and he was very angry with me. After he finished chewing me out, I called the general manager and found out his check had been mailed to an alternate address he had put on his tax forms. Turns out his check was waiting for him in a P.O box that he was in the process of closing. But I learned from this that it is my responsibility to make sure the mailing address is correct for everyone who works for me.

9

BUILD TECHNIQUES

Now it is time to talk about how to build a sound system. Modern sound systems are incredibly complex and often have thousands of pieces of equipment. It is not something you simply slap together. It takes lots of planning and preparation. It takes lots of drawings and research. It takes knowledge of how to build a system so you can visualize the system and order all of the parts and pieces. It takes an attention to details and a flexibility to know that everything will change. It takes patience, and if it is done correctly then you will have nothing to do at the shop other than slap labels on equipment and plug it together.

Paperwork

The absolute most important part of the build is the paperwork. There are some standards in the business, but not everyone's paperwork is the same. The importance of the paperwork is not in its sameness but in its completeness. Kai Harada, sound designer of *Million Dollar Quartet* and associate sound designer for *Wicked*, describes paperwork like this:

> *My personal opinion is that so long as the drawings are coherent, the data concise, and the lists clear, it really does not matter much what system one uses. As with the actual sound system, maybe it is not as big a deal what kind of loudspeakers one uses, provided they are used in the proper way.*

When you walk in the door at the shop, the first thing anyone cares about is the paperwork, and you will be bombarded with questions about it. More than likely, the shop has requested paperwork weeks before you reached the shop, and if you were able to give it to them then you will be ahead of the game.

The shop is going to want several things. They will want rack drawings so the show key can look over the system. A good show key is constantly checking your paperwork to make sure there are no problems. The shop will want your cable order and your bundle sheets, which we will talk about soon. The shop will want your perishables order. More than likely, you will walk into your zone and it will be pretty empty. You will need to order gaff tape and colored electrical tape and other build supplies that you put on your perishables order. The shop will also want to go over hardware with you. They will want a list of speaker yokes and video monitor yokes. They will want to discuss speaker towers. All of this is crucial to get the build started, because you have a limited time of two to four weeks to build the system, and if the shop does not order the equipment and start building the hardware, then you will not leave the shop with a complete system. If you walk in with completed paperwork you will be in good shape, but if you walk in with more questions than answers about the system and not much paperwork, then you are going to be in a deep hole and it won't take long for the shop to get very frustrated with you.

So what makes up good paperwork? It starts with drawings. There are two types of drawings that are important to building a system: rack drawings and signal flow drawings. Rack drawings are simple drawings that show what goes in each rack and give some important information about the equipment and sometimes the patching of the rack. Rack drawings should include a front and back drawing of the rack. Rack drawings are sometimes done in Excel but are usually done using a drafting program like Vectorworks or AutoCad or Stardraw. Rack drawings are usually drawn to be printed on an 8½" × 11" regular sheet of paper. Signal flow drawings are detailed drawings about how devices flow through the system. These drawings are overviews of the system. Usually signal flow drawings are broken up into different sections of the sound system. There could be a signal flow drawing for the inputs, outputs, video, intercom, midi, and network systems. Sometimes signal flow drawings are printed on large paper, such as 24" × 36" sheets. My personal preference is to break it up to fit on 8½" × 11" sheets of paper.

In Figure 9.1, we see a rack that is designed to go front of house at the mix position. This drawing shows the equipment that needs to live in this rack and where the design team wants the equipment to live in the rack. There are

Figure 9.1 A rack drawing.

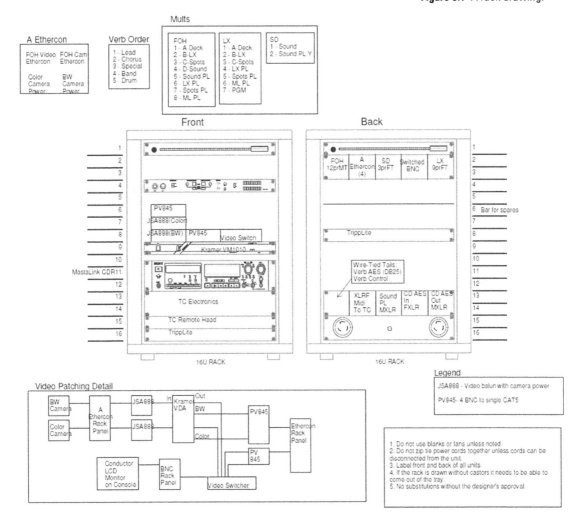

many times when you are building a show from a drawing where things need to shift once you start actually building the rack, but it is important to stay as accurate to the drawing as possible. As soon as you move something, you find out there was a specific reason that wasn't clear to you about why it had been drawn that way. This drawing also includes the mult tails that need to go in the rack. We will discuss mults later in this chapter. It also gives a breakdown of the mult tails and of the patching scheme for the video components in the rack. Rarely do drawings include every answer to every question, but the rack drawings should offer enough answers that your build crew can put the rack together and patch it as needed.

Figure 9.2 An output signal flow drawing.

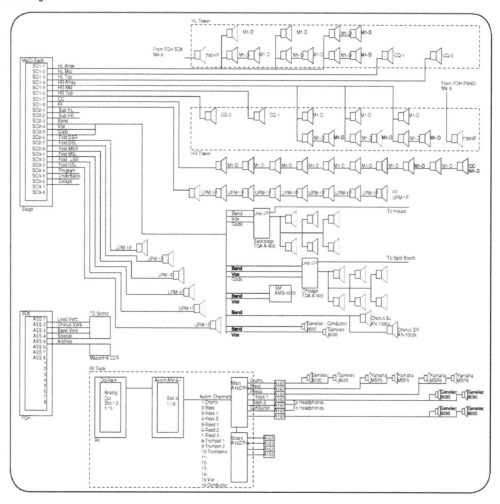

Figure 9.2 is a typical signal flow drawing. This drawing focuses on the output section of the sound system. It gives you an overview of the speakers and amps in the system as well as any processing. It is a quick way to understand how A gets to B. It is not meant to answer every question about the sound system. If you try to add too much information on the signal flow drawings, you will find the drawing growing out of control and hard to maintain. When working on a system, it is important to know where to look for information and for that information to be provided in bite-size pieces. If you add too much information to your drawings, you can actually confuse people. For example, if you put the mult line number on your rack drawing for the House Left speaker, put that on your signal flow, create a label for it, then you run the risk of it changing and not updating it in all three places. That can confuse people trying to build the system. It is better to understand what people are looking for when they look at different pieces of paperwork. When looking at a signal flow drawing, people want a general overview. They want to quickly understand the components of the system and how they flow.

Figure 9.3 A midi signal flow drawing.

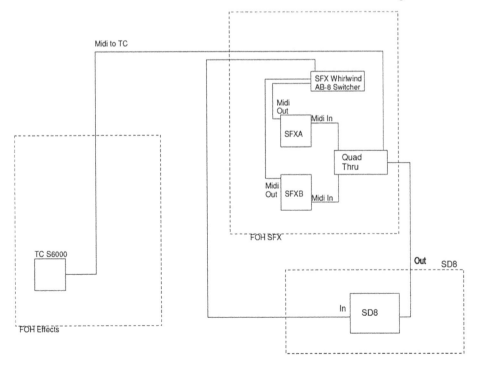

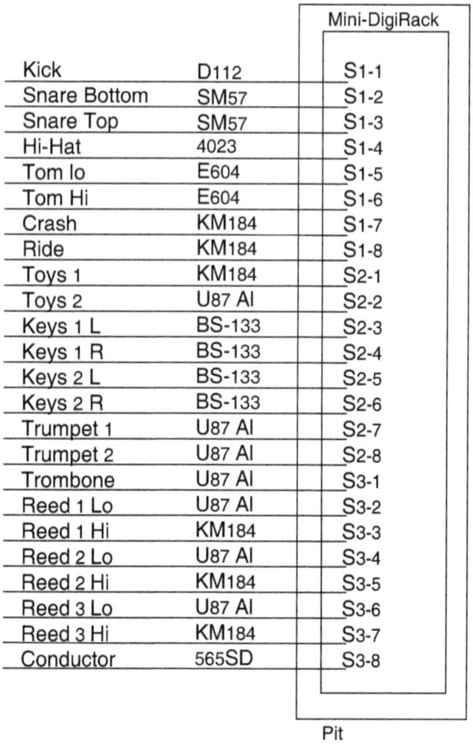

		Mini-DigiRack
Kick	D112	S1-1
Snare Bottom	SM57	S1-2
Snare Top	SM57	S1-3
Hi-Hat	4023	S1-4
Tom lo	E604	S1-5
Tom Hi	E604	S1-6
Crash	KM184	S1-7
Ride	KM184	S1-8
Toys 1	KM184	S2-1
Toys 2	U87 Al	S2-2
Keys 1 L	BS-133	S2-3
Keys 1 R	BS-133	S2-4
Keys 2 L	BS-133	S2-5
Keys 2 R	BS-133	S2-6
Trumpet 1	U87 Al	S2-7
Trumpet 2	U87 Al	S2-8
Trombone	U87 Al	S3-1
Reed 1 Lo	U87 Al	S3-2
Reed 1 Hi	KM184	S3-3
Reed 2 Lo	U87 Al	S3-4
Reed 2 Hi	KM184	S3-5
Reed 3 Lo	U87 Al	S3-6
Reed 3 Hi	KM184	S3-7
Conductor	565SD	S3-8

Pit

Figure 9.4 An input signal flow drawing.

Figure 9.3 is a signal flow drawing for the midi. The midi flow can be fairly complex on a sound system. We use midi to trigger all kinds of devices. It is important to show how the midi system works, and this is a good place to show the midi channels that different devices are transmitting to or receiving from. Figure 9.4 is a signal flow drawing for the inputs. It very simply shows what is plugged into the different consoles on the show. It shows the name of the device and the model of the device.

It is very common for designers to do a signal flow drawing of the inputs and outputs and hand that drawing off to their assistant or associate or A1; the rest of the drawings and paperwork are then done by those people. It takes a lot of knowledge of the final product needed in order to take a signal flow drawing like that and transform it into a full-blown, well-documented sound system, but once you know what you are doing, it is a good way to work. The first step in this process is to start doing rack drawings based on the flow drawings. You start putting the gear into racks, and then you start to see the need for other equipment. Before you know it, you are trying to figure out how to cable the system.

More than likely, the designer did an equipment order that was given to the shop based on the signal flow and that is what the shop based their bid on. As you work on the rest of the paperwork, you will start to make a list of the other equipment needed for the system. This includes cable, adaptors, racks, and boxes. As you build this paperwork, you will build a list of supplemental equipment that will need to be given to the shop. Once you arrive at the shop, one of the first things you will want to do is get a list from the shop of what equipment they have put in their system for your order. You will want to compare that list with your list and make sure they match. This is a very important step, because if their list doesn't match then you could get to the end of the build and be missing some important piece of gear that is now back-ordered and not available. You also want to make sure the lists match so that as things change in the shop, you can give the shop an accurate list of changes to your order.

One reason why paperwork is so important is that the systems are so large that there are times when you need to be able to trace problems in the paperwork to figure out how to fix the problem. Another reason is because this may not be the last time you build the system. If you are lucky, your show will be a big hit and go on tour, and you will need to build it again for the tour. If the paperwork is accurate, this is an easy process. If it isn't, then you end up making the same mistakes and digging yourself out of the same holes that you already worked your way out of the first time.

In the shop, several pieces of paperwork are important. First is the order from the shop, which is the full list of everything you need to build the show. The order should be available for anyone to look at. Second are the checklists. It helps to have checklists that the build crew can use to keep track of what has been done. Figure 9.5 shows a typical checklist. The checklist shows you the name of the device and the model; it also shows you the location or box of the device and the color if it is to be color-coded. There is a box to mark when the item is done. Some checklists will also have a box for "Pulled" and "Labeled" so the item can be marked off as those things are completed.

Figure 9.5 A typical checklist.

AMPLIFIERS				
Done Device Name	Model	Box (Location)		Color
Balc Fills	E-PAC	Amp Rack D		
Mezz Fills	E-PAC	Amp Rack D		
Orch Fills A	E-PAC	Amp Rack D		
Orch Fills B	E-PAC	Amp Rack D		
Balc Delays	H5000	Spares Rack		
Balc Rear	H5000	Amp Rack D		
Bass/Guitar	H5000	Amp Rack A		
Brass	H5000	Amp Rack B		

When I build a show I purchase file jackets, which are just like file folders except they are closed on the sides so it is more like a big envelope. I label the file jackets for the different parts of the system build, such as "Microphones," "Speakers," and "Bundles." I place the checklist and labels for each part in its respective file jacket. That way when the microphones are brought into our zone, I can hand someone the "Microphones" file jacket and that person can go off and label all the mics and let me know if we received everything we were supposed to receive.

When I do my paperwork for a show, I define which box or rack every device in the system goes in, and by doing that I can print a checklist and the appropriate labels for each box. I then create a file jacket for each rack, and when I assign a person to build a rack I give him the file jacket. Once he has the file jacket, he has the rack drawing, the checklist, and the labels for the rack.

Labels

Labels are important. They are crucial. They are demanded. I have worked for a couple of designers who will look around the system after it has been loaded in and if they find a cable that isn't labeled, they will unplug it. It is a good lesson to learn. The point is that your equipment should be labeled in such a way that you do not need any paperwork when you have your head buried in your doghouse, which is the term used for a box built around the back of a console to hide all the ugly connections. You should be able to pick up any cable and read what it is and where to plug it in. Without labels, there would be chaos.

The labels for the system are usually done before you get to the shop and they are printed out in the first day of the build. For labels we typically use Avery labels sizes 8167 or 5167, and 8160 or 5160. The difference between the 81 and 51 is LaserJet versus InkJet. The 8160 label is called the mailing address label. The 8167 is called the return address label. Sometimes we use 8165, which are a full 8½" × 11" size label. There are also generic versions of these labels that are identical.

Basically, with labels we are labeling two distinct things. We are labeling devices and connections. The word *device* was established for use in a sound system by the Sound Commission at USITT. A device is any item in your sound system. A microphone is a device, and so are a speaker and a rack and an XLR barrel. We want to label every device in the system so that anyone looking at a device can read what it is. That means that we want a device label on the front and back of every piece of equipment (Figure 9.6a and b), so that once you rack the gear, you will be able to see a label when you have your head in a rack looking at the back of a bunch of amps. When I print my device labels, I typically print three identical labels for each device so we have one label for the front, one for the back, and one for when you make a mistake. Device labels can be either large (8160) or small (8167) depending on the item. It is important for a device label to have the show name, the model number, and box or location of the device.

The other kind of label is for connections. Connection labels are always small 8167 size. Basically, a sound system is made up of devices with connection points, and we want to label them all. A connection label is similar to a device label but holds more information. As seen in Figure 9.7, in the upper left corner of the connection label we see the name of the connection's parent device. In the upper right corner, we see the name of the connection. This can be thought of as the signal for that connection. To the right of the label in the space in between the labels, we see the model number of the parent device. In the yellow section of the label below the name we see the patch information. This tells you exactly what the connection plugs into.

There are some rules that need to be followed when applying labels (Figures 9.8 and 9.9). These labels are paper labels with a gummy backing. The first thing we need to do is apply a piece of colored electrical tape or

Figure 9.6a A device label.

Figure 9.6b A version of a cable label done by Kai Harada.

Figure 9.7 A connection label.

Conductor A	
Equip:	SHINTRON IXS VDA
Note:	
Patch:	
Location	Pit Rack

Conductor B	
Equip:	SHINTRON IXS VDA
Note:	
Patch:	
Location	Pit Rack

Figure 9.8 A device label properly attached to a piece of equipment.

Figure 9.9 A connection label properly attached to a piece of equipment.

PVC tape to the device. Then the label is placed on top of the electrical tape. This makes it much easier for the shop to remove your labels when the gear is returned. If you put a paper label directly on a device, it will leave glue residue when you try to remove the label. It is important not to stretch the PVC tape as you apply it to the device. If the tape is stretched, it will slowly recoil and pull itself off the device. The paper label is delicate and not made to last forever. The next thing we need to do is protect the label. We use transparent tape to completely cover the label and electrical tape. This gives the appearance of a laminated label. Be careful not to buy invisible tape. The name is deceiving because it is not invisible; it is cloudy. Transparent tape is completely see-through but is more expensive.

Not all labels in your sound system will be printed paper labels like the ones above. Another important tool in the arsenal of the Broadway sound person is the Brother P-Touch. There are lots of times when you will need to P-Touch labels for devices and connections. The process is exactly the same as with the paper labels. Most commonly we use TZ-Tape for the P-Touch. The most common sizes are ½", ¾". There are times when you will need ¼" or 1" as well. Finally, the last line of defense in the battle to label your system is white gaff tape and a Sharpie. A few years ago, Sharpie started making Twin-tip Sharpies, which gives you a normal Sharpie on one end and a fine-point Sharpie on the other end. These are the bomb.

I have mentioned color coding a few times, and I want to take a minute to expand on it. Every show on Broadway has a color coding scheme for the sound system. You color code things by using different colored electrical tape under the labels. You can also use different colored gaff tape when color coding large items such as cable crates and racks. The reason for color coding is to give some instant visual understanding to your sound system. The most important things to color code are your snakes, mults, and XLRs (Figure 9.10). There are several different theories on color coding, and it is usually up to the mixer to create a color code.

The most common system, and the one I use, is to give each snake a unique color for the box it is going to live in. An

Figure 9.10 A rack color-coded for multiple mults.

example is in the doghouse of my console. Let's say I have three input snakes. I color code them patriotically as red, white, and blue. All the XLRs of snake A will be red. All the XLRs of snake B will be white. All the XLRs of snake C will be blue. This makes it easy for me to look in the doghouse and pick out the XLRs for a specific snake.

Personally, I have found this to be the best use of color in a sound system. There is also a system that all inputs are one color, all outputs are another color, and all intercom lines are a different color. I have also seen a system where interconnect cables out front are one color and stage runs are another color and pit runs are another color. It is a good idea to use color to help you. Whatever system you decide to use is fine, but just choose one and understand the goal and how easy it will be to color code your equipment using your system. I have seen some people try to color code different types of signal, which becomes a nightmare to build because you could have a snake with four different colors. That type of system will be very time consuming and complex for the people building your show.

Mults

I have mentioned mults several times, and now it is finally time to explain what a mult is. A common item in sound is a *snake*. A snake is a cable built with multiple conductors. One end of the cable usually has a breakout to several XLR tails, and the other end has a breakout to an XLR box. Basically, a mult is a snake that has a detachable box on one end and a detachable fan-out

> of XLR tails on the other end. The multicore cable that runs between the box and the tails is called a *trunk*. Mults come in a variety of sizes or pairs. Wireworks makes multichannel audio cables that are the standard for Broadway, and these multichannel audio cables are usually called mults.

Walk backstage in any Broadway house and you will find MultiBoxes and MultiTails in almost every rack, and MultiTrunks will be no doubt dangling from above and plugged into the racks. The same holds true for most theatrical tours. Wireworks multicable or multicore is so integral to Broadway that Wireworks even created the Broadway Latching System, or BLS, to address the needs of the community. So how did it all start? How did this cable come about and how did it become an industry standard?

Gerald "Jerry" Krulewicz, president, and Larry Williams, chief operations officer, started Wireworks in 1974 because they saw a need for multicable. Jerry explains,

Our original markets were live theatre and broadcast. At the time there was no such thing as multicable. People used mic cable and taped it together with friction tape. It was a ridiculous situation. So we started a company. The cable was actually borrowed from the computer industry, which used shielded cable. We started with a stage box hard-wired to a fan-out on the other end. Back in the XLR days people would use small numbers of mics, but once multicabling came along, people wanted to put mics everywhere.

When asked how Wireworks multicable was initially received, he explains,

It was a product that was released at the right time to the right market. So it was like a miracle. It was a whole revelation. It made load-ins in New York quicker and helped immensely with touring. Some of the original designers started to name the different trunk cables and some of those standards have stuck. Between Otts (Munderloh) and Tony (Meola) they came up with a naming system which has gone on to become a standard. The industry not only adopted the multicabling idea, but the whole concept. This worked very well for the rental shops as well. It eliminated their need to have custom panels for every tour.

Abe Jacob, sound designer of *Jesus Christ Superstar* and the original *A Chorus Line* among many others, says,

We first used homemade multicable or snakes on the West Coast in the late 60s. The remote recording trucks of the time, I believe, introduced snakes to the live sound world. I also remember bundling mic cables together to make our own sort of snake, but just for microphones. With multi-pin connectors and fan-outs, it became much simpler to build processing racks and it certainly became easier to run cable in mults wherever needed. Wireworks has been in the forefront of multicable design and fabrication and has been most helpful in assisting designers with new products that can serve many applications.

Tony Meola, sound designer of *Wicked* and *Steel Pier*, among others, explains how he started using Wireworks multicable:

The first show I remember using multicable on was I Remember Mama designed by Otts Munderloh. I remember when I walked into Masque and there were six three pair mults on the order. They said you'll get three XLRs taped together buddy and you'll like it. Otts really pushed for multicable. When I told him they won't give it to you, he said we'll tell them they have to. Before multicable we bundled cables together. Back then the only multicable was more of a snake without breakouts. It was a box or tails at one end and tails at the other. It wasn't easy to tour. And it was either 15 or 19 pairs. It was so long ago. I remember with Wireworks we would say can you make us a custom panel? And for every show we had custom panels. Eventually to keep the cost down I said we don't need custom panels every time, we just need three, six, and nine pairs standard.

Otts Munderloh, sound designer of *Barnum* and *Dream Girls*, to name a couple, talks of his beginning with Wireworks multicable:

Multicable made it easier on the road. New York is an anomaly. You load in and everything changes. But it made

it much easier on the road. I would have to say the tour of A Chorus Line *was the first show to use multicable. In New York it was not multi. And that tour was '76–'77. Jerry and Larry approached me to do my cables for* A Chorus Line. *I knew of Jerry. Jerry was the electrician on the bus and truck of* Promises Promises *the same year I was out on the bus and truck on* I Do! I Do! *and that was in '74. So I approached Abe and said do you want to use this? I was the mixer on* A Chorus Line *and I was mixing for Abe. That was in 1975. The shops were extremely resistant to buying multicable. One of the shops said to me, "If we buy this stuff then no one is ever going to use it again so we are going to charge the show for the price of the cable." And when I did my next show I said, "Where are my multicables?" and they said they were all rented out. And that is how it started.*

For our purposes, we use mults to connect the sound system together. We use mults to send 19 XLRs from the RF Rack to the console. It is important for you to know the basic terminology for mults and the number of pairs available. Mults come in 3, 6, 9, 12, 15, or 19 pairs. A pair is actually the three wires needed for an XLR. Of course, the mult tails could also be terminated for connectors other than XLRs. Mults have clips and ears that are used to plug the trunk into the tails. Mults can have a tail fan-out or the connectors could be built into a box that can be safely placed somewhere, or into a panel that can be built into a rack. MultTails can either be Stecked or Inline. Stecked means to attach the MultTail to a plate that allows you to mount the tailset in a rack using Steck Rails, such as a Middle Atlantic FK-2.

There are rules for labeling mults correctly, and neatness counts (Figures 9.11 through 9.16). Figure 9.14 shows a 19pr that has been labeled correctly. To label a MultTail correctly, it may be necessary to remove the clips so you can wrap the PVC tape completely around the connector. You also want to wrap the clear tape completely around. If you don't get a full wrap with the clear tape, then the clear tape will peel off and the label will be destroyed. By looking at the label we can see that it is a bundle label and the bundle is "Pit A," which is printed vertically on the right. The box this bundle lives in is the "Pit Bundles" box, which is printed vertically on the left.

Figure 9.15 shows a typical rack all wired up. If your cables are color-coded and labeled properly then it will be easy to find what you need. In Figure 9.16, the trunk name is "P.B" and the model is a "19pr 150." The "From" line is highlighted and tells us which end of the cable we are looking at. The "To" line tells us where the other end of the cable is. A trunk has a

Figure 9.11 A MultBox that has been color-coded and labeled.

Figure 9.12 A MultPanel that has been color-coded and labeled.

Figure 9.13 A 6-pair MultTrunk that has been color-coded and labeled.

Figure 9.14 A 19-pair MultTail that has been color-coded and labeled.

male and a female end, but sometimes we write "M-(Pins)" and "F-(Holes)" so there is no confusion about what end is the male and the female.

Figure 9.17 shows a labeled XLR. When labeling a connector, it is a good idea to label it without restricting the ability to open the connector. If you find a problem with a connection, you will want to open it to check the wiring. If the label has been done incorrectly, you will need to remove the label in order to check the wiring.

Figure 9.18 shows a labeled Mult-Box. A common use for MultBoxes is as sub-snakes in the pit. It is common to put a rack in a pit that has some 19pr panels that carry the mics to the console, and from the pit rack you run mults to the different sections in the pit. This picture shows the MultBox that is a sub-snake for the conductor. When labeling anything, you should think about how someone will hold the device to read it. You want the labels to be right-side up and read from left to right even when plugged in. It is also a good idea to make a decision on whether the labels on MultBoxes will be above the connection or below. Being inconsistent with that can get really confusing.

Figure 9.19 shows a labeled Mult-Panel. There is not a lot of room on MultPanels for labels, so you do the best you can. MultPanels are also not always the same. Sometimes a 19pr panel is a 2-space panel with two rows of XLRs, and sometimes a 19pr panel is a 3-space panel with three rows of XLRs.

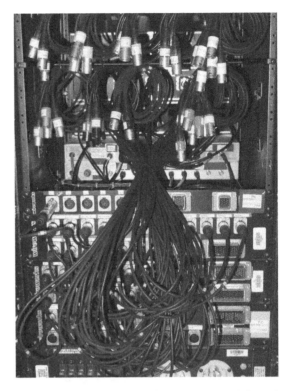

Figure 9.15 A typical distro rack filled with MultTails and MultPanels.

Figure 9.16 A labeled bundled trunk connector.

Figure 9.17 A labeled XLR connector.

Figure 9.18 A labeled MultBox.

Figure 9.20 shows a labeled Mult-Trunk. This trunk was done incorrectly. There are several problems with the way this cable was labeled. Take a look at it and see if you can find the problems.

So here are the things done wrong with that label. The first thing you should notice is it is just messy. The person who did this label did not take the clips off the connector so the tape could be wrapped around. The PVC tape is hanging past where the connector seats, so there is no way this trunk will be able to be plugged in. The clear tape is not attached to anything on the right side. This label will pull off before you get it out of the box at load-in or it would pull off as you run the cable in the theater and it gets pulled through some hole in the wall.

The way most shop preps begin is by labeling mults. This process can take a couple of days. Basically, everyone sits around a table with colored electrical tape and clear tape and razor blades and screw drivers and sheets of labels and a checklist. We will use a 3-space metal blank, lay out the colored electrical tape, and cut the tape into small pieces with the razor blades. This process always starts a little slow as people get a system going, and then it ramps up. It is a good idea to take the time to do this part correctly, and it is to be expected that someone will label an entire 19pr female tails before realizing he needed a 19pr male tails. That is where the spare labels come in handy. The challenge for the person who does the labels is that on day one all the labels need to be done for the connections in the show, which means basically that everything has already been thought out.

Figure 9.19 A labeled MultPanel.

Figure 9.20 An incorrectly labeled MultTrunk.

Bundles

Another important term in building your show is *bundles*. A bundle is several cables that will be taped together using friction tape so they act as a single run (Figure 9.21). You bundle cables together that will run the same path. A bundle cuts down on the number of runs you have to pull at the load-in. For example, you will bundle three 19prs for the pit so you can save time at the load-in and only have one run to the pit for the 57 mic lines in the pit instead of three individual runs. There are places outside of New York that call a bundle a *loom*. While this may be an acceptable term where you come from, it is important to learn and use the terminology that is used in New York. If you persist in walking around the shop talking about your "looms," you will probably have a hard time getting your bundles built.

Before you arrived at the shop, hopefully you sent the shop your bundle information. That usually includes bundle sheets, a bundle summary, and possibly a list of the cables needed to build all the bundles. Figure 9.22a shows

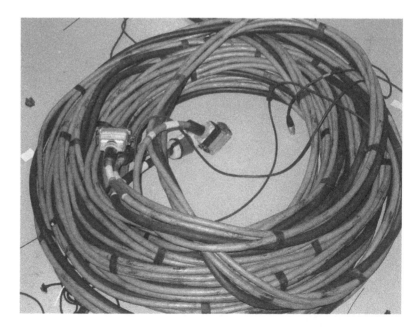

Figure 9.21 A bundle.

part of a bundle summary. A bundle summary shows all of the information about the bundles for the show. A bundle sheet (Figure 9.22b) is the exact same information as the bundle summary, but it only has one bundle per page. The bundle information on the summary and sheet shows the cables that will be bundled together, the color e-tape

FOH A

250'

Tag this end

Description:		Box:	Bundles A		From:	To:
Cable Name:	Cable:	Color:	Label:	Ampland	FOH	
FOH COM	15PR 250'	Black	Large	FROM@Ampland	TO@FOH	
N	15PR 250'	Red	Large	FROM@Ampland	TO@FOH	
Cue Light	EDISON 250'	Black	Large	FROM@Ampland	TO@FOH	
Aviom	ETHERCON 250'	Yellow	Small	FROM@Ampland	TO@FOH	
FOH	L5-20 250'	Black	Large	FROM@Ampland	TO@FOH	

FOH A Ext.

50'

Tag this end

Description:	Extensions (Bottom of box. Ears on FOH End.)	Box:	Bundles A		From:	To:
Cable Name:	Cable:	Color:	Label:	Ampland	FOH	
FOH COM	15PR 50'	Black	Large	FROM@Ampland	TO@FOH	
N	15PR 50'	Red	Large	FROM@Ampland	TO@FOH	
Cue Light	EDISON 50'	Black	Large	FROM@Ampland	TO@FOH	
Aviom	ETHERCON 50'	Yellow	Small	FROM@Ampland	TO@FOH	
FOH	L5-20 50'	Black	Large	FROM@Ampland	TO@FOH	

Figure 9.22a A bundle summary.

Foldback Trunk Ext

50 feet

Number of Cables 5

From **Galileo Rack**

_To_ **Svc Truss**

Mult Letter	Cable Type	Cable Name	Gender at Galileo Rack	Gender at Svc Truss	Pulled	Labeled
FG	6	Foldback Ext	FEMALE Clips	MALE Ears	☐	☐
VG	4CAT5	Video Truss Ext	MALE	MALE	☐	☐
	NA-C3	Foldback AC Ext	BLUE	GREY	☐	☐
	NA-C3	Video AC Ext	GREY	BLUE	☐	☐
	NL-4	Offstage PGM NL4 Ext	MALE	MALE	☐	☐

Figure 9.22b A bundle sheet done by Kai Harada.

for each cable, and where each cable is going from and to. A bundle summary is used as an overview to keep track of what has been done. The bundle sheets will stay with the cable. The normal order of a show build is to label mults first; while you are labeling mults, the shop pulls the bundled cable. Once you are done with the mult labels, you move on to the bundles. It is a good idea to get the mults and bundles done first, because it takes a long time to get this stuff labeled and bundled.

The shop pulls the cable for your bundles. Sometimes the shop guys will lay the bundles out on the floor in a stack for each bundle. The bundle sheet will be placed on top of each stack. Then your build crew will label the bundles. As with everything, there are rules on how to label bundled cable. The rules on how to label a mult apply to bundled cables, but there are a couple of additional rules. A bundled cable has a different label for each end of the cable. If you look back at Figure 9.16, you will see how the "From" line is highlighted. That means this label is meant to go on the male end of the cable. The female end will look the same, except the "To" row will be highlighted. When you label these cables, it is necessary to tag one end of each cable to show which ends are to be on one end of the bundle. You are tagging one end so the guys in the shop can easily tell which ends group together.

Typically, we tag the end with an obnoxiously colored gaff tape, such as green, pink, or orange. We use these colors because they are very obvious. It is also common to tag the "To" end (Figure 9.23). The "To" end is the end of the cable that will be pulled to a location. For example, if you are doing a pit bundle and the cable is going to run from the Pit to FOH (Front of House) and the box at load-in is going to be in the pit, then you would want the "To" end to be the FOH end because you are pulling the cable to FOH. If you tag these cables in a way so the gaff tape is permanent, then this tag can have a second use. When you open a box, see a bundle, and see the bright green tape on the ends, then you know that end pulls out of the box and to its destination. If you do not see the tape, then you know that the bundle was put in upside down and must be flipped over.

Figure 9.23 A bundle with the cables tagged for the "To" end.

Communications

When we talk about communications, we are talking about walkie talkies, intercom, and wireless intercom. We are talking about voodoo and the bane of our existence as sound people. We are talking about the most important and least important thing we have to do. It is the most important because without communications, or com, there can be no show, but it is also the least important because it has no effect on the quality of the sound of our show. Be that as it may, we are responsible for com and we have to learn everything we can about it to make sure it works flawlessly.

On Broadway, we use many channels of com to keep conversations private when needed and public when necessary. The typical com system utilizes eight channels of com. The normal breakdown is Deck, Lighting, Spots, Sound, Lighting Private, Spots Private, Moving Lights Private, Sound Private. The stage manager gets the four "public" channels, which are Deck, Lighting, Spots, and Sound. This allows the stage manager to have communication with all the departments involved in running the show. The Deck channel consists of carpenters, flymen, props crew, and other stage managers. The Deck channel is also connected to the wireless com system. There are times when the wireless com system has two channels. In this case, the people on wireless can have a private channel.

The conductor normally gets a single channel, which would be on the Sound channel. The sound crew has a public channel and a private channel. The sound crew may have several different places onstage for com. It is common for there to be several places for the sound crew to listen to the wireless mics, and it is common to place a com beltpack at each of these positions. Mixers have different needs. Some mixers only want the Sound channel and the Sound Private channel, but others want all of the public channels at the mix position.

The lighting department is the most demanding when it comes to com. Usually each member of the lighting team gets a four-channel remote station with the com channels laid out like this: A—Lighting Public, B—Lighting Private, C—Spots Private, D—Moving Lights Private. Often the lighting people will call the Lighting Public channel the SM (or stage manager) channel. To them, the only reason they talk on that channel is to talk to the stage manager. The spot operators normally get a two-channel beltpack with the Spot Public and the Spot Private channels. There are times when one of the spot ops will call all of the spot cues. It is possible that only one spot op gets a two-channel beltpack and the other spot ops get a single-channel with just the Spot Public channel.

The reason the lighting channels are broken up this way is because different people need to have private conversations with different people at the same time. The lighting designer will mostly be talking to the moving light programmer and will occasionally talk to the stage manager. One of the assistant LDs will spend all of his/her time talking privately to the spots to explain what their cues are. Another assistant may be spending most of his/her time talking privately to the stage manager to make sure the cues are where they are supposed to be.

Recently, projection design has become very popular in theater. With the rise in the popularity of projection comes a need for more com channels. It may be necessary to add a Public and a Private channel for the projection team. Often the projection team will want the Lighting Private and Lighting Public channels as well. Lighting and projection have to work closely together so they are constantly talking with each other.

When planning a com system, you will need to reach out to the other design teams to find out their needs. Figure 9.24

Figure 9.24 A signal flow ICOM drawing that represents a standard Broadway com system.

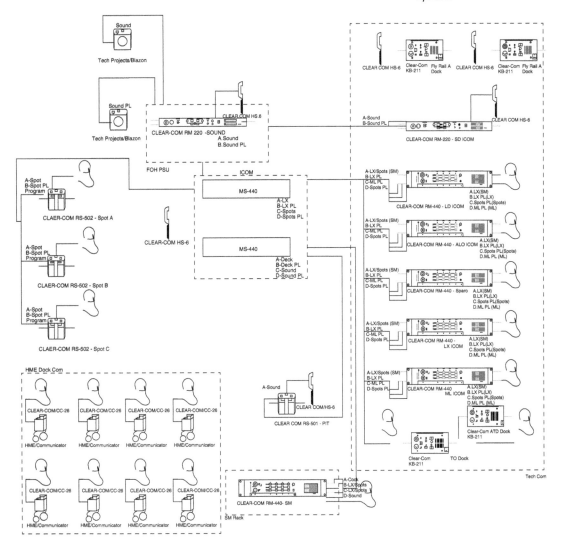

shows a typical intercom signal flow. You also have to remember that your com system is temporary. It must be easy to set up in the house and easy to dismantle. It also needs to be easy to expand. You will be asked to add beltpacks for spot op subs, training stage managers, and light board ops to learn the show. You will also be asked at some point to create a calling recording of the show, which is a recording of the show on the left channel and a recording of the stage manager calling the show on the right. In order to make this recording, you can use a Clear-Com DC Blocker, which is an inline XLR barrel, to strip the 38 volts off the com line and plug it into your recording device on the right channel. We are not allowed to make recordings of the show unless requested for training purposes.

Video

We are also responsible for the CCTV, or closed-circuit TV, for the show. This video is not artistic and is not meant for the audience. This is video that is crucial to the running of the show. The most common camera needs are Conductor camera, FOH Black and White Low-light camera, FOH Color Camera. The Conductor camera is a shot of the conductor and is distributed to anyone who needs to follow the conductor. This shot will go to the mix position and the stage manager, as well as to a couple of video monitors on the balcony rail. The video monitors on the balcony rail will need to get their power from lighting and go through a blackout generator, which will allow lighting to cue these monitors on and off instantly. The conductor shot will also go to any location for offstage singing. The conductor shot may also be distributed throughout the pit. It is common to put a 5.6" LCD, or Delvcam, on top of most, if not all, music stands in the pit. The drummer and percussionist may require several conductor shots in their area.

The FOH Black and White Low-light camera allows the stage manager to see the stage with the lights up or down. We hang IR, or infrared, emitters that allow the cameras to see in the dark. This helps the stage manager call the show better, because he or she can see when things are placed in the dark. We also have an FOH Color camera because sometimes the black and white just isn't high enough quality for the stage manager to call the show. These two cameras are normally hung on the balcony rail. Most of the cameras we use are typical CCTV cameras that are used in the security industry, but recently the FOH Color cameras being spec'ed are professional HD video cameras. The Conductor camera is usually a small bullet or lipstick camera. Figure 9.25 shows a basic video signal flow.

Figure 9.25 A signal flow video drawing that represents a standard Broadway video system.

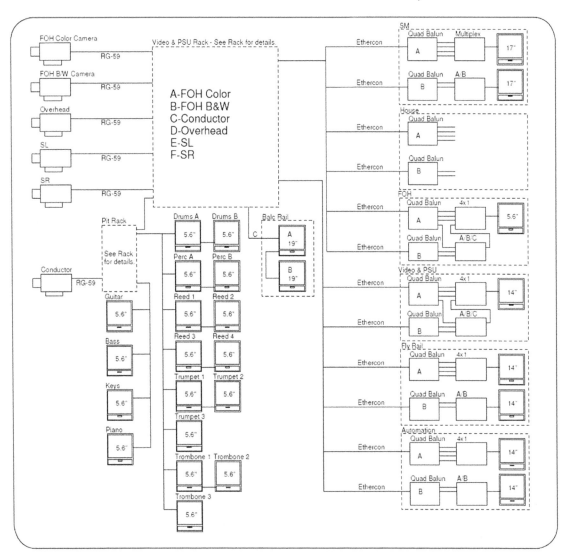

It is also possible for the show to need several other cameras. These are specialty cameras needed for stage management and automation to call the show. It is common to have an overhead camera on one of the electrics. It is also common to hang a camera that is able to be controlled remotely. Either the stage manager or the automation person will run the remote. These specialty shots can't always be predicted. It is a good idea to leave room for expansion in your video system.

Testing

Before we leave the shop, we put the entire sound system together and test it. This is a crucial part of having a smooth tech. I worked on one Broadway production that was an absolute mess. Not a single person on the design team had ever worked on Broadway; they had done very little planning for the build, and unfortunately the planning they had done was mostly wrong. Building that show was like pulling teeth. No one had an answer to any question. There were no bundle sheets. There were very few labels. The drawings were wrong. It took us three weeks in the shop to build the show and on the last day in the shop, we were still building bundles and labeling mult tails. The build was so messed up that nothing was tested in the shop. The result was a load-in that was painful. We would run a bundle out and find that one cable was backwards in the bundle. We would go to run in a speaker cable and find that cable was never ordered for that speaker. This had a snowball effect and led to a miserable tech. We were constantly behind. There was always something that didn't work or wasn't planned for. Even though this happens occasionally, it is not the norm and it is not the goal.

As things progress with the system build, you will want to start planning for the testing period. You will want to plot out a space for the FOH and the Ampland. It is a good idea to have a drawing of what these areas will look like. Then you place the racks exactly the way they will be in the theater.

Then all the bundles and cables are patched in, and the system comes to life. It is the most exciting time in the shop. After weeks of planning and labeling, you finally get to see what works and what doesn't.

When we test the system (Figures 9.26 through 9.30), we patch the entire system together before we start. This can make for a messy build zone because it takes so much cable to patch a system together. Usually we will start testing on either the video system or the com system. To test the video system, we set up the cameras and hang a piece of paper in front of the camera that has the name of the camera shot written on it. That allows us to go to all the monitors in the system and verify the camera shots are correct. Next, we test the com. This involves moving from com station to com station and making sure all channels work. The signal flow drawings come in very handy when testing the com and video systems. Those drawings make it easy to see if you have tested everything. When testing com, it is a good idea to first test the call light of each channel, then test the sound of each channel. There is nothing worse than hearing feedback from an unterminated

Figure 9.26 Scott Armstrong and Justin Stasiw setting up for testing.

Figure 9.27 Scott Armstrong patching some backstage racks.

Figure 9.28 Ampland and the monitor desk setup for testing.

Figure 9.29 FOH setup for testing.

Figure 9.30 Racks patched for testing.

Clear-Com channel. To avoid that, you can start with the call light. If the call light is not responding normally, then there is a problem and the channel will probably squeal if you turn it on.

After we check the com and video, we usually move on to testing the paging system. Testing the paging system includes testing the God mics, which are the mics used by the director, choreographer, and stage manager. Once this has been tested, we are usually in a good place to dismantle parts of the system to be packed up. While that happens, we test the inputs and outputs of the system. Part of testing the outputs includes listening to the speakers. The speakers we use are rental speakers, which means they could be at different places in their lifecycle. It is a good idea to listen to your speakers to make sure the components sound the same. Sometimes the shop will even test the speakers for you using Smaart or some other acoustical measuring program. Also, we will test the speakers for polarity using a Cricket or a Minilyzer, to make sure the amps and speakers are wired correctly. To do this, we send a popping sound to the speaker using one end of the Cricket, and we use the microphone on the other end of the Cricket to make sure the popping coming out of the speaker is polarity correct. If you don't test for polarity, you could end up in the theater with speakers cancelling each other out and a very unhappy designer.

Packing

Once all of the testing is done, the only thing left to do is to pack up the system. All of the equipment needs to be safely packed so it can travel on a truck to the theater. The process of packing a system can take days, which is why we start packing the video and com system as soon as possible. Before the testing is done, a lot of things can be packed up. The pit mics and stands can all be packed. The speaker yokes and spares can all be packed. Hopefully, a lot will be packed up before the testing is done.

When we pack a system, we tend to organize the system so it can be easily be found. We will pack all of the FOH (Front Of House) bundles in one or two boxes and label the boxes something like "Main Runs." We try not to put "FOH" on a box unless the box itself is meant to be placed front of house, so in labeling the FOH runs we avoid the term "FOH." We will box all of the pit runs in a couple of boxes and label them. We will pack things up that go to the same location in the theater and we will label the box with the location the box needs to go at load-in. We also pack similar speakers together. Intercom will be its own box. Video monitors will take up several boxes.

We will pack a workbox with adaptors and video cameras and fragile equipment. We also pack similar cables together. There will usually be a box of NL4 and a box of power cables and a box of XLR. This doesn't always work out perfectly. I did a load-in recently in which the pack was done in a strange way and it was challenging to find what you needed. In the end, it was my fault for not watching how things were being packed more carefully. If that happens, you do the best you can to re-organize at the theater.

Packing is important for several reasons. One is that this gear is rented, and if it is damaged the production will have to pay for the damage, so when we pack we wrap delicate items in bubble-wrap and use foam to protect items from banging around. Part of packing involves understanding how things travel. Packing your system on a truck is like a living, horizontal game of Tetris. One big difference is that if you get your pack wrong and have to redo it, you are going to endure a slew of ridicule from the Teamsters and stagehands who are loading the truck. When you pack a truck, you need to know how wide your truck is and how many boxes will fit in to create a solid wall. You need to know what can stack and what can be flipped "wheels to the sky." You need to know where to put load bars, which are solid bars that hook to the sides of the truck to create a stopping point so the gear can't move. You also need to know where to put ratchet straps, which are like load bars but are thick straps instead of solid bars. The people loading the truck will take care of most of this, but it is a good idea to have a decent clue before you get near a truck.

3

THE TAKE-IN OR LOAD-IN

This is where the rubber meets the road. All the planning led to this moment, and now you get to see what you got right and what you got wrong. Hopefully, you planned for surprises because there will definitely be plenty of those. If you are lucky, all the design teams communicated well and no one will be shocked when you say you are flying a center cluster array. Hopefully, no one will be stunned that you have speakers that need to hang on the electrics. Hopefully, the cable runs will be the length you planned for. And so the fun begins.

Before you get to the theater, there is usually a week of pre-hang. This is where the carpenters come in and hang motors. It is possible that some lighting has been hung as well. It is also possible for some speakers to be hung during this period. If any speakers are going to be hung or sound cable is going to be run, it is because it needs to be done before something blocks the path. If that is the case, the house electrician or the tech will let you know. As the mixer, you should check with these people about the pre-hang. You should also check with the production electrician before the load-in starts to make sure there are no surprises between you and the rest of the electrics department.

The house electrician is a very important person once the load-in starts. Every Broadway theater has a house electrician. The house electrician is the boss of the electrics department at the theater. He or she can make the final call on lots of things at the load-in. As sound people, we are considered to be in the electrics department and are sometimes listed as assistant electricians. There really is no negative to this. It actually makes sense. Most of the work we do intertwines with lighting, so it stands to reason that we would be in the electrics department. The house electrician is responsible for hiring the crew at the theater to load the show in and to run the show onstage.

As the A1, you are considered the head of your department, even though your department is a subset of the electrics department, and as such you will work with the house electrician to make sure you have enough crew to do what you need to do. You will also need to make sure the house electrician knows what speakers you are hanging and how you are going to hang them. Most Broadway houses are historical landmarks or are treated as such. The theater owners usually have an opinion about how things look in the theater, and it is up to the house electrician to make sure you don't do something that goes against what the theater owner wants.

Even though you are considered to be in the electrics department and the head of your department, you do not always work for the house electrician. In many cases the mixer is hired by the production and put on a special contract. This allows the production to be the mixer's real boss. Other times you are hired by the house electrician after you have been approved by the designer and production. In this case, the house electrician is the mixer's real boss. It can be a little confusing, but it is best to always assume that both are your boss and give both the respect they deserve.

The main difference between the two comes down to money and power. You usually make less when you are hired by the production and, if the production is unhappy with the job you are doing, they can fire you fairly easily because you are an at-will employee. If you work for the production, you are on what is called a "pink contract," which is a contract issued by I.A.T.S.E, the International Alliance of Theatrical Stage Employees. If you are working on a "pink contract," you are sometimes referred to as "road crew" or "production crew" by the house crew. Every Broadway production is allowed to have a certain number of pink contracts. The production does not have to have any pinks or contracts, but if they choose to there are agreements between the union and the producers about the maximum number of contracts allowed on a show.

Many times the producers prefer to have the sound mixer be on a pink contract instead of working for the house because they feel it gives them more control over the mixer and the sound of the show. Some sound designers feel the same way. If the designer and/or producer are unhappy with a contract mixer, they can easily fire that mixer and hire a new one. If the mixer works for the house, the designer and producer will have to talk to the house electrician about replacing the mixer. Even though I can't imagine a situation where the production was unhappy with a house mixer and the house electrician refused to replace the mixer, it still adds a level of complexity to the process. Another reason is that if the mixer works for the house, then the house electrician will have to approve any mixer who might come in to sub or replace the main mixer. Again, this is just another level of complexity.

Also, as mixers we are required to go to press events for the show. This morning I was at *Good Morning America* at 5 am for the show I am currently mixing. If the mixer is on contract, then the production tells the mixer they need him to go to an event and arrange a fee for the event, but if the mixer works for the house this will go through the house electrician and the rate is non-negotiable. But probably the main reason to prefer to have the mixer on a pink is to save money. The contract crew on a show typically makes less than the house crew, which means most shows want as many contracts as they can get on a show. The opinion as to which is better, a house job or a contract job, is a matter of perspective. They are both jobs and they both pay well. One is not necessarily better than the other. I have worked both types many, many times. For the most part, I am treated the same either way, and if contract jobs had not been available when I started mixing on Broadway, I don't know if I would be mixing today.

10

DESIGN TEAM'S JOB

The designer definitely gets more involved at the load-in than he may have at the build. The designer may stop by occasionally, but don't expect him to stay there all day. The designer's assistant and possibly his associate will be at load-in the entire time to answer questions and make sure things are progressing the way they should. It is more than likely that the designer is working on another show and doesn't have a lot of time to be at the load-in. It is common for designers to set goals so they can be around for important events. The mixer is typically the conduit of information about how the load-in is going, and it is important for the mixer to be able to assess the status and predict how long projects will take. A designer might tell the mixer that he wants to be around when the center array flies out, and he may ask when it will fly out. It is up to the mixer to analyze the load-in and plan a schedule with goals and to meet those goals. The last thing you want is for the designer to show up because you told him the array would be flown and it isn't even close.

The designer will also be there to focus the speakers and make adjustments to the system. The system drawings and paperwork are a good jumping-off point, but it is very likely that speakers will need to be adjusted once you see them in the space. If there are any issues with sound equipment and lighting or scenic elements, the design team can communicate with the other design teams to come to a compromise. The sound design team also works behind the scenes with wardrobe and wigs to develop a mic plan for how the actors will be mic'ed. If you are lucky, all the women will be wearing wigs and the wig designer will allow you to put the mic packs in the wigs. This is very common on Broadway.

There are many benefits to putting the mics in the wigs, but the main benefit is that the mics last much longer. The mics tend to go through less stress in the wig and the connectors can last for months, if not years. Typically, when we put the mic packs in the wig we order a shorter 18" version of the mic. If you are going to put the pack in the wig, you need to plan this out in advance with the wig department so that when they build the wigs they leave room for the pack. When the packs are placed in the wigs, the process starts with a wig cap on the actress. Then the pack is wrapped in the wig cap and pinned into place. Then the mic is pinned into position and placed exactly where the sound designer wants it. Most actresses really like the mic pack in their wigs. It is something they never have to think about; there is no cord running down their backs and no tape slapped on their necks.

The design team will also be dealing with sound cues and recordings for the show. Hopefully, before tech starts, the sound cues for the show will basically be built and loaded into SFX or some other playback system. There are times when shows use click tracks on Broadway. It is not very common and it is a closely guarded secret when it is used. A click track is a recording of vocals or other sounds played back with the orchestra. It is called a click track because it is a multiple channel audio cue that has a metronome on one track for the conductor and orchestra. Typically, the click track is fired by the conductor and the band plays along to the metronome. In the house, the recorded vocals are mixed in with the live vocals, if there are any live vocals.

Actors' Equity has to approve the use of any click tracks, and they are reluctant to approve these recordings. For Actors' Equity to approve the recording, it would have to be proven that the use of the track would not negate a job. A show that has a big dance number and the cast is too winded to sing while dancing isn't necessarily going to be approved to use a click track, or vocal sweetener. Actors' Equity would recommend hiring more actors and having them sing offstage instead. If, however, the song involves the two leads in the show and it isn't physically possible for them to sing while doing the action in the scene, then Actors' Equity would probably approve the use of the click track.

There are times when a conductor will use a metronome with no vocal track. In this case, no approval is needed because it is solely for the musicians. *Hairspray* is a good example of this. In order to keep the music the same tempo for every show, the conductor and musical supervisor decided to use a metronome. Every musician in the pit wore headphones and listened to the metronome while they played. The result was a perfectly timed show every time.

The design team will also coordinate with the music department on the pit layout and instrumentation, as well as plan for special needs in the pit. The music department consists of several people: the composer, the orchestrator, the arranger, the copyist, the music supervisor, the conductor, and the music contractor. Some of these positions will be obvious and familiar. The composer obviously is composing the music. Once the composer has composed the basic structure of the music, the orchestrator takes over and fleshes the music out for a full orchestra. On a revival where the original composer is dead, a show will have an orchestrator deal with needed changes in the incidental music and tweaks to the score. The orchestrator's job on a revival is to maintain the intent of the original orchestration while bringing a new point of view to it. The arranger deals with the vocal arrangements. The copyist documents all of the music and prints it out for the musicians. The music supervisor is basically the director for the musicians. The music supervisor is similar to a music director. The music supervisor can make decisions about instrumentation or tone of the music and works closely with the orchestrator to create his or her unique vision of the music in the show. The conductor conducts the orchestra. The conductor's job is to maintain the vision of the music supervisor. The music contractor hires the musicians. If you want to play in a pit on Broadway, you will want to get to know the music contractors.

The design team will communicate with the music supervisor to find out his likes and dislikes. Ultimately it is the sound designer's job to help bring the music supervisor's vision to life. If you like more guitar than the music supervisor, then you are going to have to learn to like less guitar. There have been times when I have said to a music supervisor, "Was the guitar too quiet in that song?" and I get the response, "No. It was perfect." And I begrudgingly reply, "Oh, good. I thought that's what you wanted." There are lots of things that need to be figured out with the music supervisor. Some supervisors really like Aviom mixers and headphones for their musicians. Some hate it. Some like having monitor speakers all over the pit. Some want none. Some like tons of baffling and Clear Sonex, which is a treated Plexiglas. Others want none. It is important to remember that it is their pit and we are there to accommodate their needs.

The sound designer will also be expecting to have a couple of sessions of quiet time. Quiet time is time when everyone else leaves the theater and the sound people get funky. Well, maybe not funky. It is when we turn on the pink noise and blast it for hours and EQ the room. Different designers have different techniques for EQ'ing a room. Some will use Smaart. Some will use Meyer Sound SIMM. Others will use their ears. And a few will unfortunately propagate the stereotype by using Steely Dan. Typically, we are given the theater in the evening to do this work.

The quiet time sessions typically occur in the days just before cast onstage, which is the first day the cast will wear mics and be onstage. When the designer does his quiet time, he will want several mics around the theater so he can compare the sound in different parts of the theater. He will also want someone dedicated to moving the mics around. It is important to document the seats where the mics were during the EQ'ing, so that if the designer needs to go back for some reason then it can be done accurately. Once the system is EQ'ed, the load-in is done and it is time to tech. There might still be things to be done, but the system has to be up and working and stable at this point.

11

MIXER'S JOB

It is the job of the mixer to be the leader at the load-in. The mixer is expected to have the answers and if he doesn't, then it is his job to get them as quickly as possible. It is not the job of the mixer to physically load the show in. A crew is hired to load the show in. It is the mixer's job to stay one or two steps ahead of the load-in crew so that, as they finish one task, another is lined up and ready for them to start. Usually the load-in crew is not part of the build crew, so they have no knowledge of the system that is being loaded-in. It is always a good idea to have a house person on the show build, but it doesn't always happen. If you have a house person on the build, he or she can keep you from making big mistakes. Even if you have a house person, though, mistakes will have been made. As the mixer, it is your job to correct the mistakes and to keep the load-in progressing.

Inevitably, all of the planning you did and all the paperwork done will end up being changed. It is common to get to the theater and find out the designer wants the speakers hung in a different place, and that means your bundles are now all wrong and your amp racks are wired wrong for the new placement. It is all just part of the process. As the mixer, you have to know your system well enough to know how to adjust it quickly for any changes and do it gracefully. It is a bad idea to get bogged down in the minutia of loading-in. It is enough to tell your crew to run out some bundles and explain to them the destination of the bundles. You do not need to get too involved with the path of the cable runs or how they are going to run them or how many people it will take. There will be one person on your crew who is considered the house sound person. If you communicate your needs to this person, you can let that person do the work for you.

It is much more important for you to make sure the crew can find the equipment it needs and the equipment is in the building. It is important for the mixer to communicate with the shop and have the shop send down equipment needed in a timely manner so it doesn't slow down the load-in. The mixer also needs to communicate with the house electrician to explain the plan for the load-in and the schedule and goals. The house electrician needs to know that you need two extra people to do something specific so he can call in the people for you. He also needs to know when the designer wants to do quiet time. He will also want to know when you think you can cut down your crew.

The mixer will spend a lot of time working on the FOH setup. This is their world and most mixers are very particular about how it is set up. I know of one mixer who likes to build all of his FOH racks in 12u racks so he can sit in a normal chair while mixing, while most other mixers put their consoles on 16u racks and prefer to stand and mix or sit in an engineering chair to mix. My preference is to have a higher mix position so I can choose to stand or sit, depending on how I feel. Some mixers also like to sit lower because it keeps their ears in the same plane as the audience. There really is no right or wrong, but there are times when the theater dictates the height. Some theaters would make it impossible to sit in a normal chair to mix and see over the audience, while others would require it to keep the mixer from being too close to the ceiling.

The next issue is about where to put the script. Older analog desks, like the Cadac, had the capability to add a sliding stand for the script. This made it easy for designers to move the script out of the way when they needed to work on the desk. More modern digital consoles make this much harder. On almost every show I have worked on this is a major discussion, and I have yet to see a perfect implementation of a script on a digital console. The DiGiCo SD8 is pretty good because it actually has a shelf area for the script, but that is the only desk that has thought about the problem. Over the past few years I have started using the script in electronic form on a computer. I place a 17" LCD monitor on an arm that can be kept out of the way and then I place a wireless numeric keypad on the console that can flip through the pages in the script. The keypad is very small and leaves the console open for the designer.

The mixer will need to bring some tools to the load-in. Some mixers have large work boxes filled with tools and are ready for anything, while other mixers are minimalists and bring a limited assortment of tools. Some mixers have their own Porta-Band Saw to cut Schedule 20 pipe to size, while others will borrow this from a carpenter. The same goes for hammer drills and other power tools. It is a good idea to have some tools, but the mixer is not necessarily responsible for having every tool. After all, we are not carpenters. A mixer is expected to have the basics, though.

Tools are bulky and heavy. When you work a gig and you have to transport your tools, it is important to know what tools you will really need so you can carry as little as possible. There is no sense in carrying an impact drill unless you really need it. A torque wrench is great to have, but not always crucial. Some tools are must-haves that we all know about: a Leatherman or a Gerber, a flashlight, a headlight, dikes, a knife, gloves, a Sharpie. There are other tools that are just as critical but not as common. There are three items that have lived in my backpack for the last ten years. I never leave home without them and I never do a gig without them. These three tools are so small that I don't even notice I am carrying them, and I can slip them in my pocket on a gig and carry them around all day without feeling weighted down. As far as I am concerned, these are three of the most essential tools you can have in your toolkit. The first is the Noise Plug by GTC Industries. The second is the Tone Plug by GTC Industries. The third is the SoundPlug by Vizear.

There's a moment at every gig where something doesn't work. You load-in all morning and set up 20 microphones for the band and some wireless mics, and you go to do a scratch test and inevitably the first thing you have is someone's scratch doesn't work. This is a perfect time to pull out the Noise Plug. The Noise Plug is a pink noise generator built into an XLR connector. It has a red LED on the top and it runs on phantom power so you do not need a battery. You can use it to check your lines. Plug in the Noise Plug and turn on phantom power and you will see a red light, which tells you the line is good. If the line is bad, you can trace back through your signal flow until you find the problem. You can also use the Noise Plug's pink noise generator to test your system if you are using a desk that does not have a built-in pink noise generator.

Now that your inputs are working, it is time to deal with your outputs. This is where the SoundPlug comes into play. If you send noise down a line and hear nothing, then pull out the SoundPlug. The SoundPlug is a piezo speaker built into an XLR connector. The top has a soft rubber opening that fits nicely into your ear. You can plug the SoundPlug into the output line and stick the SoundPlug in your ear and listen for your signal. The SoundPlug does not require batteries and comes in very handy when someone shows up at the last minute and wants an audio feed for his camera and insists he is not getting signal. Just plug the SoundPlug in and if that person hears the signal, then the problem is on his end.

After checking your inputs and outputs, it is time to do some basic system testing, which means it is time for the Tone Plug. The Tone Plug looks just like the Noise Plug, except it has a small button that selects between 11 functions. The Tone Plug has five sine wave test tones at frequencies of 100, 250, 400, 1 k, and 10 kHz. It also includes a 40 to 2400 Hz signal, and a short multi-frequency pulse for adjusting speaker time delays, reverbs, and echo units. There is also an amplitude sweep function that helps you adjust compressors and limiters. The Tone Plug is truly an amazing sound system multi-tool. Throw these three plugs in your bag and take them with you everywhere.

12

UNIONS

All Broadway theaters are union houses, and if you are going to work on Broadway, you will need to be in a union. There are many different unions involved in theater, and they break down by departments. There is a union for actors and directors and stage managers. There is a union for musicians. There is a union for company managers. There is a union for stagehands, and there is even a union for ushers and box office staff. Some of these unions you might be familiar with. Others are more specialized for New York theater. These unions are all considered craft or trade unions, not industrial unions. (Craft unionism means to organize a group of workers, unifying the workers in a particular industry along the lines of the particular craft or trade they work in. Industrial unionism is when all workers in the same industry are organized into the same union, regardless of differences in skill.)

Craft unionism formed the backbone of the American Federation of Labor, which merged with the industrial unions of the Congress of Industrial Organizations to form the AFL-CIO. Under this approach, each union is organized according to the craft, or specific work function, of its members. For example, in the theater trades, all stagehands belong to the stagehands' union, the musicians belong to the musicians' union, and so on. Each craft union has its own administration, its own policies, its own collective bargaining agreements, and its own union halls. A union hall is a place where workers can go when they need work. Theaters call the hall when they need workers.

There is a lot of union pride on Broadway, and there is a long history of the different unions working together for the betterment of all workers on Broadway.

Local One

Local One is the Stagehands' Union in New York City. The history of Local One goes back to 1863, when the first stage employees' organization was founded at the home of Brother James Timoney in New York City. Originally named the Theatrical Workman's Council, its name was changed to the Theatrical Mechanical Association. On December 26, 1865, the T.M.A. was incorporated under a state legislative act titled "An Act for the Incorporation of Benevolent, Charitable, Scientific and Missionary Societies." The T.M.A was a good start for theater labor, but it wasn't perfect.

On April 26, 1886, 41 T.M.A. members met at 187 Bowery, where a new charter was drawn up to create the Theatrical Protective Union Number One. The new charter's preamble to the constitution states:

We, the theatrical employees of the theatres of New York, deem it eminently right that we should organize for the development and improvement of our conditions, asking but a fair and just compensation commensurate with the service rendered so that equity may be maintained and the welfare of our organization promoted, accepting any wise, honorable, and conservative mediation as a proper adjustment of all difficulties that may arise.

In the early days, it was not easy to prove a need for decent wages because there were plenty of unskilled workers willing to work every night to handle the scenery so they could watch the shows. Back then the shows were fairly simplistic. As shows became more complicated and stock companies were replaced by traveling companies, the acceptable standards for the work done by stagehands increased. That led to the need for skilled craftsmen, which is what drove the cheap men and free amusement workers from the field.

Vaudeville became a major employer of T.P.U. members, and T.P.U. members helped protect the material of one act from being stolen by another act. Stagehands were very vocal if they caught someone stealing material from another act. The act would either drop the questioned material or they would have to deal with stagehands during their performance. Many theater traditions were formed during this time. Scenery was changed by a whistle cue: the stage mechanic would blow his whistle to signal commands to the other workers and drops would fly in, scenery would move, and props would clear, which is the reason whistling in the theater is considered bad luck. In 1916, T.P.U. Local One began to put its union stamp (the "bug") on all scenery and equipment in its jurisdiction (Figure 12.1). Most scenery was being built in New York and sent out on the road.

Figure 12.1 A Local One bug.

Today Local One is a union of over 3,000 workers. Local One's jurisdiction is New York City and it covers workers in theater and television. Every Broadway theater is a union theater and hires union crew. The way it breaks down in New York is that the theaters are owned by companies such as Jujamcyn or the Nederlanders. These theater owners rent their theaters to producers. The theater owners and producers belong to the Broadway League. The Broadway League negotiates a contract with Local One, which defines the pay rate among work rules for the Broadway theaters. The theater owners hire the house heads (house electrician, house carpenter, and house props). When a producer rents the theater, that producer agrees to abide by the contract accepted by the League and the union. Then the house heads hire crew to fill the calls. The union has a business agent who oversees the process to make sure the rules are followed and helps to settle disputes between the crew and management. Every theater elects a steward to oversee the process and deal with issues on a local level.

There are two ways to gain membership to Local One. The first is to apply to be in the apprentice program. If you pass the test and are accepted, you will be assigned to work as an apprentice at a Local One house or shop for three years. The other way is to make a certain amount of money for a certain number of consecutive years on Local One payroll. Currently the requirements are three consecutive years of making a minimum of $35,000. The only way to make this money is to be hired by a house head or a Local One shop.

I.A.T.S.E.

I.A.T.S.E. is the International Alliance of Theatrical Stage Employees. I.A.T.S.E. is the union that formed around the same time as Local One. Local One is part of I.A.T.S.E. and there are Locals all over the country. The Locals are numbered in order of their inception. New York was first, and then came Chicago (Local 2) and Pittsburgh (Local 3). Boston is Local 11 and Local 300 is Saskatoon. My good friend Scott Armstrong toured for years and he made a shirt of a fake Local. It was Local 666, the venue from Hell. It was a very popular shirt among stagehands. When you tour it is very common to get SWAG (Stuff We All Get) from the local crew. Most of the time you buy the shirts, but sometimes they are gifts. After you finish a tour, you end up with dozens of shirts proudly displaying the Local number and the city name with some logo. Fort Worth's shirt has a long-horn's skull decked out in stage lighting. Several Locals use the logo that simply says, "Bad Stagehand. No Donut." Flint, Michigan has the saying, "Remember when you were a dick to me on the in?"

I.A.T.S.E. covers theater, television, and projectionists, and film crews in many cities. It is the parent for all of the Locals around the world. Just like New York theaters are union houses, so are most of the road houses around the country. A road house is a theater that mostly does touring productions. In those houses, the house crew are all union employees. There are non-union road houses as well, but most of the theaters are union.

When you tour as a mixer you work on a special touring contract. It is called a "pink" because it is on pink paper. The pink contract, which is the same as the pink contract used on Broadway, is a contract that is collectively bargained for between I.A.T.S.E. and the Broadway League. The Broadway League is also comprised of theater owners and presenters from around the U.S. The pink can also be called a road contract; when you go out on a pink, you are considered the road crew and not the house crew. It is very similar to the way it works in New York. When you work on a union touring show, it is called a "yellow card" show or a "legitimate" show. It is called a "yellow card" show because I.A.T.S.E. gives you a yellow card to prove you are in the union. At most venues the local crew will meet the local business agent, who will want to see their card and contract. This is standard practice to confirm the validity of the show as a legitimate show.

It is possible for shows to tour without using union crews. If this happens, then when the show plays a union house, there could be rules the road crew has to follow. Since that crew is not in the union, they could be asked not to touch anything. If you find yourself in this position, it is best to follow the rules and be polite. Remember that it is their house and you are a guest. It is also possible for a show to tour with a union crew and non-union actors and non-union musicians. This happens because different unions have different policies and it usually happens on tours that are doing one-nighters, which involves travel and a show on the same day. I.A.T.S.E. has found a way to work with producers of these types of tours and has started unionizing them. Actors' Equity has not been able to work this out yet.

When working in New York as a mixer, the majority of the time the mixers are hired by the production and put on a pink contract. That means the mixer works for the production and not the house. It also means that the mixer is paid according to the pink contract and not the Local One, or house, contract. The pink contract has a minimum weekly that must be paid. That weekly is extremely low for New York, which leaves the mixer to negotiate for himself to get a decent rate. Also, by having the mixer on a pink, it puts the mixer in a position at the theater where he is the head of his sub-department and yet not really in charge because he is not in the hierarchical structure of the house crew.

Most of the time this works out with no problems, but occasionally it can pose problems when the mixer wants something done and the house crew is not obliged to follow the request. But this is rare. I did a load-in in Canada one time on a non-union tour and the local crew was not happy about us being there. I can remember looking at the house head sound and saying, "The next thing we normally do is hang these speakers," and he looked right back at me and said, "We're not going to do that right now." But for the most part the system works extremely well and everyone is respectful.

802

Local 802 is the largest local union of professional musicians in the world and is comprised of members who live or work in New York City and Nassau and Suffolk counties, New York. Their mission is to fight for the interests and well-being of the musicians employed in New York's music and entertainment industries. Local 802 is the collective bargaining agent for musicians working on Broadway. Their members include the orchestral musicians who perform at Lincoln Center, the Broadway pit musicians, the freelance musicians who perform in the many other musical venues in and around New York, the jazz artists who perform in our city's world-famous jazz clubs, hotel, club date and cabaret musicians, recording musicians, orchestrators, arrangers, and copyists, as well as teaching artists and the musicians who work in the rock, blues and contemporary music scenes.

On Broadway, 802 is the backbone of musical theater. Without these talented musicians, there would be no musicals. To establish a contract, 802 bargains with the League. Part of the contract establishes a minimum number of musicians allowed per venue. The larger Broadway houses have larger musician minimums. That means if you are going to do a show in the Palace, which is one of the largest Broadway houses, then you are going to be required to have at least 24 musicians. If you are going into the Helen Hayes, which is a smaller house, the minimum is going to be much lower. The assumption is that if you are going into a bigger house, then you are a bigger show with a bigger budget. It is possible to petition the union for a smaller minimum if the show does not require that many musicians. *Title of Show* is an example of a musical that only required a small band and petitioned the union for and was granted the reduced minimum.

In 2003 the League and 802 were having trouble coming to terms on a contract. The League was pushing to cut the minimums in the larger houses from 26 down to 14. When 802 was not willing to accept that cut, negotiations broke down and 802 went on strike. The League had been preparing for this for the entire year prior. On March 7, 2003 the musicians went out on strike and the League brought in non-union musicians to rehearse every show on Broadway with computer-based virtual orchestras. This consisted of one musician playing a keyboard that sounded like a 26-piece keyboard band and a conductor. The entire cast and crew was brought in to do a full rehearsal with the virtual orchestras with a scheduled performance that night. I was mixing *Man of La Mancha* at that time. It was by far the worst sound ever heard on Broadway during that rehearsal.

As the day went on it was surreal. The actors were stunned at what they were going to have to perform to. Everyone was respectful of the situation but uncomfortable. Outside the theater, our musician friends were picketing, and inside, we were listening to a disaster. Around 4 pm, Local One announced that they would be honoring the picket line, which meant that Local One stagehands would not cross the line to do the show. At that point, everything stopped in the theater. The house electrician shut down all of the lights and sound and the props crew helped the virtual orchestra people take their equipment out of the theater. Between 4 pm and 5 pm there was confusion. The League did not want to give up and Actors' Equity and I.A.T.S.E. had not honored the line, so as an actor or a pink contract person we were still required to show up for the show. Luckily, Actors' Equity announced at 5 pm that they would honor the line as well. With that, the shows were cancelled and we all joined our musician friends walking the picket line.

That is why we are in a union. We look out for each other and make sure we have a safe working environment and can earn a decent wage. In 2007, Local One went on strike and the actors and musicians marched the picket lines with the stagehands; so did ushers and company managers. An injury to one is an injury to all.

Here is a press release that Local One put out about the strike:

Dear Brothers and Sisters of Actor's Equity, Musicians Local 802, Operating Engineers Local 30, Teamsters Local 817, ATPAM, Local 306, Local 751, Local 764, Local 798, Local 829 and Local 32BJ:

This past Tuesday evening at curtain time, the producers put out a press release announcing that they intended to implement onerous work rules on Local One. An hour later, they backed off, sending a second release saying they would implement on Monday, October 22.

Local One found out about the producers' latest moves when Mayor Michael Bloomberg called James J. Claffey, Jr., President of Local One, to offer his help, which the union respectfully declined. Nobody at the League of American Theatres and Producers had the courtesy to call the Local One President.

From moment to moment, no one seems to know which work rules the League intends to implement from their expansive list of demands. That's what Local One has been dealing with during these negotiations. It started three years ago when the League verbally threatened Local One at the table, while also initiating an assessment on every ticket sold to the public to create a $20 million war chest to break our union.

Some additional background: on September 7, after only five introductory meetings and on the day hard bargaining was to begin, Bernard Plum, the League's lead negotiator, declared that September 30 would be a day of reckoning. On October 9, the League presented its final offer to Local One and the press at the same time. Their final offer was not written for Local One, but for the media.

Ignoring the League's deadline, Local One put its entire book on the table and, as Local One President James J. Claffey, Jr. has declared publicly and privately, the Union addressed nearly every item on the producers' list and offered imaginative solutions that met the producers' requests.

We are professionals and unashamed to state that we are defending good middle-class jobs that pay our mortgages, feed our families and allow our children to attend good schools.

The producers' numbers, so widely distributed, are misleading at best and often bogus.

Their press release celebrated an offer of 16.5% increase in wages. But the producers failed to mention their offer was accompanied by a 38% cut in jobs and income.

We are the caretakers of the theatre, the protectors of the workplace. We keep it safe for all of us. Six days a week, sometimes seven, we are the first to arrive and last to leave.

The producers' attack on flymen is ignorant to the basic safety concerns in any theatre. Without a flyman, who would be addressing safety problems over head? Who would be checking rigging eight times a week? Who is the first line of defense against any fire in the fly space?

Why do you think there are still fire hooks and extinguishers, *by law*, located on the fly floors? And, if there were not a flyman on the grid, how long do you think it would take for someone on the stage to reach that fire-fighting equipment?

Automation? We've long embraced it. Local One is more productive thanks to automation. We've modernized along with the newest technology. We build, install, manage, and repair all of it. We operate safely tons of scenery moving around in the dark and at breakneck speed without injury to you or us.

The producers will also fail to tell you and the press that Local One labor over the last few decades remains 8% of the overall cost of producing a Broadway show. We get raises only when negotiated, but the producers raise ticket prices with every new hit, not to mention $450 premium pricing.

The attack on the working professionals of Local One by the League now and you later is all about profit (although they only put losses in their recent press releases).

Last year, the League announced Broadway box office grosses of $939 million. Secret is the income from licensing, secondary rights, film rights and the hugely lucrative merchandise sales.

The biggest secret of all is the producers' real profits.

In these negotiations, we put everything on the table except the safety of the stage crew and everyone entering the theatre. The producers' attack on minimums is an attack on the safety and efficiency of the load-in of shows. It is also an insulting failure to recognize the size, the scope and the technical difficulty of the work we perform and the industry that is our life.

We stand ready to resume negotiations at any time and we stand ready to defend ourselves from the implementation of unsafe, unsound and unacceptable work rules that the producers are threatening to enact.

We are Local One. We are all under attack.

Respectfully and Fraternally,

The Membership of Local No. One

922 and 829

I.A.T.S.E. also represents designers. Local 922 was established to collectively bargain the League on behalf of sound designers. Abe Jacob helped secure a charter for sound designers within I.A.T.S.E. (Local 922); and in 1993 he helped the Local achieve its first collective bargaining agreement with the League of American Theatres and Producers, which became the Broadway League. And in 1999, he helped merge the sound designers with Local One. Sound designers were the only designers who were represented by Local One. Local One sound designers have been transferred to USA 829 for all future bargaining.

USA 829 is a union of scenic artists that was founded in 1897 as the United Scenic Artists Association. This union was briefly a local of I.A.T.S.E. until the AFL-CIO ruled that the local must leave the IATSE and join the Brotherhood of Painters, Decorators, and Paper Hangers of America (later to become IBPAT, the International Brotherhood of Painters and Allied Trades). United Scenic Artists of America Local 829 (officially, it is United Scenic Artist Local USA 829, but everyone refers to it as USA Local 829) grew to include scenic, costume and lighting designers, mural and diorama artists, scene painters, production designers and art directors, commercial costume stylists, storyboard artists, and, most recently, computer artists, art department coordinators, sound designers, and projection designers working in all areas of the entertainment industry. On April 27 of 1999, the membership of United Scenic Artists Local 829 voted by an overwhelming majority to re-affiliate with the International Alliance of Theatrical Stage Employees (I.A.T.S.E.) and to disaffiliate from the IBPAT. Sound designers joined Local 829 in 2003 for regional theater representation.

Article I. Theatrical Wardrobe Union Local 764

The Theatrical Wardrobe Union (TWU) Local 764 has represented wardrobe personnel in the New York area since it was founded in 1919 and chartered under the American Federation of Labor as Theatrical Wardrobe Attendants Union #16770. On August 1, 1942, the union was granted a charter by the International Alliance of Theatrical Stage Employees and Moving Picture Machine Operators of the United States and Canada (I.A.T.S.E.) and became Local #764 of the International. In October 1982, a revised charter was issued in the name of Theatrical Wardrobe Union. Local 764 has contracts in all areas of the entertainment industry. Its members work as wardrobe supervisors (or costumers) and assistants on feature films, pilots, soap operas, commercials, and a variety of television programs. They are wardrobe supervisors, assistants, and dressers at venues, including Broadway theaters, the Metropolitan Opera, Lincoln Center Theatres, Brooklyn Academy of Music, Madison Square Garden, Nassau Coliseum, and Radio City Music Hall.

The physical jurisdiction is defined as within a 50-mile radius of Columbus Circle for film and New York City, Long Island, and Westchester County for theater.

Equity

Actors' Equity Association (AEA or Equity) was founded in 1913 as the labor union for actors and stage managers in the United

States. Equity currently represents more than 48,000 members. Anyone who has worked in theater should be aware of Equity. Most, if not all, regional theaters use almost entirely Equity actors, and even smaller theaters use some Equity actors. Equity is probably the most recognized entertainment union.

13

THE SCHEDULE

The mixer has to be keenly aware of the schedule for the load-in. The typical Broadway load-in is three weeks. Usually the work days are Monday through Friday from 8 am until 5 pm. Even though that seems like a long time, Broadway shows are huge and get more complicated every year, and time slips away very quickly. If the mixer does not stay on top of it, there can be major problems. The first thing that must be found out is when the last day in the shop is and make sure you are ready to put the system on the truck. It is most often the case that you will truck the system down to the theater in several trucks over the course of several days. To accomplish this, you have to plan out what you will need when load-in starts and will have to pack your boxes accordingly. You will also need to mark the boxes in the shop in such a way that it is very clear what is to be sent down on what day. Sometimes we color code boxes for different days, as well as labeling the boxes with the day they are to be sent. We also line up the boxes in the shop to coincide with the trucking schedule.

You have to coordinate the trucking with the production manager. On Broadway, we use Teamsters to load and unload trucks. Teamsters are a union of workers who load and unload trucks and containers. This union is not specific to the entertainment industry. If you have a truck show up unscheduled, then there will be no Teamsters to unload the truck. The production manager is going to want to schedule the trucks to get the most benefit out of the Teamsters. The production manager, as an example, will want a sound truck at 8 am, a lighting truck at 8 am, and a scenic truck at 9 am to best utilize the truck loaders. Once your equipment arrives at the theater, you need to place it in the theater in logical places. The equipment for FOH should all be marked to go to FOH, and the stuff for Ampland, which is just a term used for the place backstage where most of the racks live, should be labeled to go to Ampland.

At this point, the dates that matter are lighting focus, dry tech, quiet time, cast onstage, pit seating, orchestra sound check, first preview, and opening night. Lighting focus gives you the date on which you have to have the com system up and running. This is when the lighting design team is going to start programming cues. Dry tech is the day when carpenters and automation are going to work through the cues in the show. By this day you will need wireless com working for the deck crew and the video system working so the automation operator can see what he needs from his console to be able to safely move the scenery. The next milestone is quiet time, which is when the designer will need the entire system on the output side to be working. The designer will also want to hear God mics and paging mics as well as the RF mics.

Probably a day or two later will be cast onstage, which is the first day the cast will be putting on the microphones. The actors typically come in at 1 pm to start this day, and the first day will either be an 8 out of 10 or a 10 out of 12, which means 8 hours of ACTOR rehearsal in a 10-hour block, or 10 hours of ACTOR rehearsal in a 12-hour block. Once you get to this stage, the schedule for an 8 out of 10 is: workcall from 8 am to 12 pm; lunch from 1 pm to 2 pm; afternoon session from 1 pm to 5 pm; dinner from 5 pm to 6 pm; workcall from 6 pm to 7 pm; evening session from 7 pm to 11 pm; notes session 11 pm–? A 10 out of 12 moves lunch from 11 am to 12 pm and actors in at noon. It also extends the evening session until 12 am. To be ready for cast onstage, you need a fully functional system except for the pit. You will need the stage manager station to be fully functioning with com, video, and paging. You will also need some way of mic'ing the actors for rehearsal. Typically, we use elastic string to create a halo for the mic, which allows the actors to quickly put the mic on. As rehearsal progresses we will move away from the halos, but this is a slow process. You will also need a temporary conductor camera setup as well as rehearsal mics for the piano and other instruments, which are usually not in the pit but in the house, so that the musical team can easily converse with everyone.

Even though you are now working between 15 and 17 hours a day, your load-in is not done. The next milestone is the pit seating. This will take place from 8 am to 12 pm on a day during the workcall. This is the day when all of the musicians will go into the pit for the first time and fight for space and let you know what their needs are. To be ready for this, hopefully the music supervisor has been able to give the sound department a pit layout of where the musicians will need to sit. The seating is what we call the session where the musicians come in and get seated in the pit for the first time. This is when we find out if there is enough space for the musicians and if they have special requirements. Before the seating, the props department will place chairs and music stands in the pit. The carpentry department has installed a platform for the conductor. The electrics department will run power in for the music stand lights so they can control them from the light board. The sound department will run in any cables it can for the mic drop boxes, sub-snakes, the conductor camera, and video monitors, and also any Aviom or speaker runs.

The seating is run by the music contractor. The sound designer and music supervisor talk to the contractor about how they would like to load the pit. Loading the pit is what it is called when the musicians move to their seats. This is usually done by sections. The reeds section will load into the pit and talk through their needs, and then they will leave the pit and the next section will move in. After all sections have gone through this, the entire orchestra loads into the pit; they play a few songs to hear what the pit sounds like and to find out if they can see the conductor. The musicians are not always mic'ed for the seating.

Next up is the orchestra sound check. This is typically the day after the seating, which gives the sound department one day to mic up the pit. This is an all hands on deck day. You get one shot at this, and if you have mics not working, the designer will be less than pleased. Normally you get two four-hour sessions with the orchestra. After that, it is time to hear the orchestra with the actors. It can't be overstated enough that mistakes are not allowed on Broadway. If you get to the orchestra sound check and something isn't working, then you are going to have a room full of people looking at you and annoyed that you are wasting their time. This is true for every step in this process. Nothing can be tested too many times. Every mic should be scratched to make sure it comes up at the board where it is supposed to. You don't want to have the Overhead Left mic coming up where the designer expects the Overhead Right.

Once the orchestra sound check is over, the clock is ticking down to the first preview. Then opening night. Then sleep. Well, after a big party. Then a hangover. Then hopefully a normal showcall schedule for the next 20 years.

TECH

The end of the load-in or take-in period and the beginning of the tech period are marked by the first day of cast onstage. The two names are interchangeable, but the term "take-in" is how it is worded in the Local One contract. The distinction between these two times is very important contractually. There are specific requirements in the Local One contract as to how long the take-in is and how many pink contracts will be allowed on the show. If the take-in is short, then fewer pinks are allowed on the production. Also, there are different rules about the responsibilities of the heads based on the length of the take-in. Once we get to the tech, we finally start to get to work on the artistic part of theater, but that doesn't mean the load-in is completely done. It is possible that there is still a lot to do during tech to finish up the show's load-in. But as far as the contract is concerned, the take-in is done.

On Broadway, there are usually about three weeks for the load-in, then three weeks for tech, and then three weeks of previews before opening. Before tech starts, everyone will be invited to the rehearsal space to see a run-through of the show. At some point during the load-in, everyone will be invited to the orchestra rehearsal space for a musical run-through or sitzprobe of the show.

14

TONY MEOLA ON MIXING

ON MY OWN, PRETENDING HE'S BESIDE ME.
ALL ALONE, I WALK WITH HIM TILL MORNING.

How many times did my fingers do everything they could to make sure the audience experienced the full power of those words from arguably the most beautiful song in a score filled with beautiful songs, *Les Misérables?* Not that my fingers were tripping along the keys of a piano, or clutching the bow of a violin, or strumming the strings of a guitar.

My fingers were on the sound console. Just about my favorite place in the world.

The only drawback to being a successful Broadway designer for 25 years is that I rarely get a chance to mix anymore. I oversee the mix. I note the mix. I even get to choose the mixer. But that's not the same thrill as being the person at the board, knowing that the slightest adjustment of any one of a number of knobs will subtly—yet greatly—impact the audience's enjoyment of what they're hearing.

A smart performer and a sensitive mixer is an unbeatable combination. How I loved giving that almost imperceptible bump in volume at the climax of a big dance number so the lead dancer could hear a little more, thus giving her the boost she needed to bring the number home. On those nights when that dancer (I'm thinking of the late Deborah Henry as Cassie in the international tour of *A Chorus Line*) would shoot me a conspiratorial look—"Thanks, Tony!"—while never breaking character . . . well, those are the moments I'd hoped to one day have before I went into show business.

Back then, I didn't even know that a thing called sound design existed. I knew I'd work behind the scenes, but I wasn't sure in what capacity. As soon as I stood at a console for the first time and realized that my love for and knowledge of music could be merged with my facility for equipment and electronics, I was hooked. Little did I appreciate that I was setting my sights on the most elusive of the four major design elements. After all, the scenery is onstage at its mark, or it isn't. Clear. The lights are on and at their appropriate color, or they're not. Clear. The actor is wearing a costume and it's not inside out. Clear. But when the level for the leading lady's microphone is in its proper level for her 11 o'clock spot . . . not so clear. Except to the mixer and (hopefully) the designer. Everyone else? Well, sound is nothing if not subjective. What sounds great to your ears might sound terrible to someone else's. Lesson number one: TRUST YOUR EARS.

That's why those lyrics from *Les Miz* mean so much to me: on my own. That's how the mixer sometimes feels; I know from experience. And I imagine that's how many a performer feels when s/he is in the spotlight. But the truth is, the two are linked: what the one is singing or saying, the other makes sure can be heard. I remember mixing Frances Ruffelle, the original Eponine in *Les Miz* on Broadway. Maybe it's because *Les Miz* was the last show I officially mixed before making the transition to sound designer. Or maybe it's because Frances and I collaborated so well. I remember discussing with her the poetic nature of the lyrics to "On My Own," which employed sensual, visual imagery that Frances conveyed beautifully. The next night, I did my part, and when she got to the lyric:

IN THE RAIN, THE PAVEMENT SHINES LIKE SILVER.
ALL THE LIGHTS ARE MISTY IN THE RIVER

I touched-up the reverb on her s's and t's. The resulting shimmer in her voice worked in tandem with her interpretation of the song. The audience response was even wilder than usual. And Frances knew that she was not on her own. I was beside her, albeit at the back of the house, and very proud to be an integral part of presenting our show as artfully as possible.

Of course, not all performers are Frances Ruffelle. In fact, few are. Less secure singers often have inconsistent, sometimes erratic responses to what they perceive to be nightly changes in sound quality. I say "perceived," because the actor is usually in the worst position to judge what we're hearing in the front of the house, which is quite different from what he's hearing—especially when there are no vocal monitors onstage. So, in addition to striving for consistency in the mix, it's crucial that a mixer understand that part of his job is to liaise with the performers; regular visits backstage for some face time with your cast will result in their having more confidence onstage. Makes sense, doesn't it? If the cast knows you care, that you are—at every moment— "with" them, that you love their performances and your job, they will assume you have their "ears," if not their back. Psychology 101, to be sure—but then a good mixer should study human nature as rigorously as he studies the latest model of microphone.

Which brings me to Shannon's wonderful book. Because all the poetry, the politics, the partnerships that made mixing a show something I loved doing—none of it would've been possible unless I understood *technically* what a console is, what an orchestration is, what a sound effect is, etc. It's like studying an instrument: You can't play a symphony until you've mastered the scales and all the other technical requirements of your particular instrument.

My instrument is the console. I've been lucky enough to play symphonies on it. The information in this book will help you do the same; the love in your heart for what theater can be will help you derive from it the kind of joy I've been lucky enough to know.

—**Tony Meola**

15

DESIGN TEAM'S JOB

This is where the designer gets busy. From this point on, the designer will be around most of the time. The designer's job is to help shape the sound of the show into what the director and music supervisor want. His job is to work with the mixer on EQ'ing the system and the mics so they sound natural. His job is to build sound cues or have them built. It is not uncommon for the designer to have someone on his team build the sound cues or even have them built by someone not at the theater. A designer on a musical is constantly looking for why things don't sound natural and trying to fix it. He is working a couple of steps ahead of everyone else to ensure things go smoothly. A designer might realize that a scene sounds bad because two actors are just too close to each other and can ask the director to position them slightly differently to make the scene sound better. He can work with the dresser of the lead actors to find a good mic placement they can help maintain. He can look ahead to figure out how to deal with the problem a hat will bring.

On *Wicked*, there was a big challenge with getting Idina Menzel to sound good when she wore the witch's hat. After trying different placements and EQs, it became obvious that the problem required more action. Tony Meola, the sound designer, worked with the costume designer to find a way to put a mic pack in the witch's hat and have the mic run to the brim. That completely fixed the sound and made the end of Act I sound infinitely better. When Tony does a show, he has a tradition of giving out shirts with the show logo and some sort of nod to the sound system of the show. For *Wicked,* the shirt had the normal *Wicked* logo with the addition of a black dot on the brim of the Wicked Witch's hat.

Probably the most important job of the sound designer in tech is to create a sound at the mix position that properly represents the sound throughout the house. For the mixer to be able to mix the show, it is crucial that he or she has a good reference point to mix from. If the show sounds perfect in the mixer's chair but is ear-bleedingly loud in the mezzanine or desperately quiet in the middle of the house, then there are going to be problems. This is something that is completely out of control for the mixer. There is no way the mixer can know how the show sounds in the rest of the house. All the mixer can do is to make it sound good in his chair and it is up to the designer to match that sound in the rest of the house. I have worked on shows where the designer has a speaker hung specifically pointed at the mixer so the designer can replicate the sound in the rest of the house for the mixer.

It is not always realistic to make the mix position sound perfect. In that case, the designer has to explain to the mixer what the difference is in his chair. I have had designers walk the house with an SPL meter and come back and explain to me that my position is 3dB quieter than the rest of the house. If the designer can find a way, it is a good idea to have the mixer walk the house to hear what the rest of the room sounds like. As a mixer, you have to be able to merge what you hear with what the audience is hearing and sometimes mix something that sounds wrong to you because you know it is right everywhere else in the house. Low-end is a perfect example of this. Almost every mix position in the country is in a bass trap. When you are mixing, you will feel like there is way too much low-end in your system. Then you walk ten feet in front of the console and realize you can't hear the bass at all. It is up to the designer to help the mixer understand what the pocket is and to define the parameters for staying in that pocket.

It may also be necessary for the designer to analyze his mixer and trick his mixer into mixing the show a certain way. If the mixer constantly under-mixes the vocals, which means to mix them in a way that they are not loud enough, then the designer may need to adjust the system so the vocals are quieter at the mixer's chair than anywhere else, which will cause the mixer to mix the vocals a little hotter and thus balance the sound in the house better. It may be that the mixer mixes the band too loud and the designer needs to make the band louder at the mix position to get him to mix the band quieter. A good sound designer will be overly concerned with the sound at the mix position. A good designer will try to get the SPL as close as possible at the mix position as it is everywhere in the house.

As I stated earlier, the designer's job is to create an environment in which the mixer can succeed. If a great mixer is working with a designer who can't EQ or balance a system, there is no way the show will sound good. If a great designer is working with a less experienced mixer, then the designer can analyze that mixer's strengths and weaknesses and do what he needs to do to get the mixer to mix a better-sounding show. The reality is that a musical can only sound as good as it is mixed. A good sound designer will know that and work with the skill level of the mixer to achieve the highest quality sound. It may be necessary to fire the mixer if he really can't cut it, but this is never a good situation. There is just no easy way to transition from one mixer to another.

Several years ago, I designed a national tour of a show that toured the country for three years. The mixer who was hired was not a very experienced mixer. As we got into tech, I realized he didn't have the chops to mix the show the way it should be mixed. We were having tons of missed pickups and sloppy mix levels. I had to sit down and look at this mixer's skill level and adjust how the show was mixed. I simplified the mix from 12 DCAs to 8. Instead of having Men and Women on separate faders I made it All. I did anything I could to simplify how the show was mixed. The result was a much better sounding show than I started tech with. As the mixer toured the show, he became a better mixer and started moving the mix back to what I had wanted, which was a bit more complicated but allowed for an even better sound. I consider that my job when I am a designer, and I know a producer wants to hear a good sounding show more than my excuses. When we were working on this show, the producer asked me what the problem was with the mixer and I said it was a hard show to mix. I have been friends with this producer for years, and he looked at me and said, "You say that about every show." I looked at him and said, "Every show is hard to mix if it is mixed correctly."

16

MIXER'S JOB

This is where the rubber meets the road. The system is loaded in and now you have to figure out how to mix the show. No matter how hard you think it has been getting to this point, the easy part is now behind you. Now you have to mix a show on very little sleep, and you will have to apologize for every missed pickup. If you miss too many pickups, the designer will find himself pulled aside to ask what the problem is with the mixer. Most directors have a very low tolerance for missed pickups. They can forgive a horrible mix as long as the correct mics were always open. This is, of course, a major challenge when you are learning the show and the timing of the actors. It is at this point where the mixer has to size up what the designer expects. Usually a conversation about what the designer considers the mixer's responsibility is a good idea. There are some designers who want no input from the mixer. This type of designer does not want to know what the mixer thinks about the sound and does not want any advice on what to do. This type of designer just wants the mixer to mix the DCAs (or VCAs or Control Groups). Some designers want input from the mixer but prefer to make the final decisions. This type of designer wants the mixer to tell him what he thinks of a certain EQ or reverb. This type of designer, similarly, wants the mixer to just mix, but he expects the mixer to give him feedback about how it is to mix on the system. This type of designer will stand close to the mixer and work as his extra hands. The mixer can say, "I need more gain on Steve." Or, "Can you work on the hat EQ on Tom during this next scene?" Then the designer can work with the mixer to make adjustments while the mixer focuses on not missing pickups.

There are some designers who prefer to let the mixer take care of gain and EQ. This type of designer will EQ and balance the system and then leave it to the mixer to make adjustments to individual actors.

It is crucial to know what the designer expects. The last thing you want to do is to start adjusting EQs if you are working with a designer who thinks EQ is off limits to the mixer. You also don't want to avoid touching the EQ even if it sounds awful because you are waiting for the designer to fix it. Once you add the band in, you have to find out the expectations for the band as well. Some designers may expect one thing when it comes to voices and something completely different when it comes to instruments. It is very common for designers to work with mixers to find a range for them to mix in. The designer may ask you to mix the vocals at −10 during the dialogue scenes. While you do that, the designer can make adjustments to the Matrix settings to allow −10 to be the level the designer wants. From that point on, it is easy for the designer to talk to the mixer about level. The same holds true for the music.

When you are going through tech, it is a test of your focus. Your job is to throw faders and not miss pickups, no matter what is going on around you. It is a hard concept to master. As you are mixing, you may have two or three people working on the board at the same time. They could be grabbing knobs and pushing buttons and doing who knows what. Your job is to focus on the actors. If a designer needs something from you, or for you to get out of the way, he will tell you. More than anything, the designer wants to shape the sound while you mix with a consistent mix. I have worked with many young mixers, and it takes them time to understand that I will let them know if I need something and to ignore me. Inevitably, I will go to grab an EQ and the mixer will pull his hands off the desk and stop mixing. Rule number one as a mixer is that you never take your hands off the faders. Never. After I repeatedly say, "Keep mixing, ignore me," the mixer will finally get it and just mix so I can work.

Script

The script is by far the most important tool to a mixer. Not a drill. Not a crescent wrench. Not a Cricket. A script is king. Without a script you have no clue what is going to happen. If you are lucky, you will have seen a run-through before tech started so you will be sight-mixing. I consider mixing to be similar to playing an instrument, and sight-mixing is just like sight-reading music. Of course, sight-reading music is easier because the notes are always in the same place and the music has a standardized look, so you have a shot. With a musical you have to program your instrument so the correct mics show up and are turned on at the right time. Without an accurate script, there is no chance of being able to mix a show.

There are a lot of strong opinions about script mixing in this business. Everyone knows it is necessary, but a lot of macho bravado goes along with not mixing with a script. There are mixers who get "off-book" as fast as they can and take pride in the fact that they were mixing without a script after the first run-through. I have worked with mixers who will give you a hard time if you mix with a script. I have also worked with some great mixers who have nothing to prove and will mix with a script in front of them for years. The reality is that the audience doesn't care if you mix with a script or not, as long as it sounds good. The director doesn't care either. In fact, no one other than sound people cares. Some sound people are convinced they mix better without a script. I think that is a very valid point. I know there are scenes I have mixed that never worked until I got off-book on them. But I am not a fan of insisting that people mix better one way over another, or by demanding that people get off-book. This is something you will have to figure out for yourself. I have worked with people who insisted I get off-book, and I had a miserable time learning the show, and I have worked with people who didn't care and I had no problem learning the show. What is important is to find out what you need as a mixer and don't let people tell you what is the right way, because there is no one right way on this topic.

One very good reason for using a script is to document the mix so it is repeatable. Personally, I don't see any other way to document a mix. At one time I subbed on four Broadway musicals. I would mix between six and eight shows a week on four different shows. When I learn to sub on a show, I am meticulous with my script. I don't push a button or move a fader unless it is notated in my script. I also make level notations for almost every line in the show. Beside a character name I will put a note that says something like "@—10" to let me know that is approximately where I should throw the fader. I think that is crucial for a sub to be able to come in once or twice a week and mix the show consistently with the way it is always mixed. When I was bouncing around these four shows, there was no way I could've remembered some of the details to the mix of these shows without the scripts. This is not to say that I do not alter the mix if needed. One of the big fears non-script mixers have is that you are just going to mix by the numbers. That is not the case. There are a lot of constants in mixing. I know that a song starts at −10 and bumps to 0. I know a reverb needs to be preset to −7.5. There is a lot I can document and follow when I mix. If I don't document that, I may end up drifting from those settings, which are what the designer set. But I can always listen to the show and realize that there is a sub trumpet player that is screaming loud and adjust. Also, if I set a song at −10 week after week for months and then all of a sudden one night −10 is really quiet, then I know something is wrong.

I once got into a debate about script-mixing with British sound designer Mick Potter. Mick designed *Bombay Dreams* and *Lady in White* on Broadway. I was the sub mixer on *Bombay Dreams*. Mick also re-designed *The Phantom of the Opera*, and I was hired to build the system, install it, and supervise the changeover and the sound of the show after it re-opened. One day we were standing in the Majestic Theater at the new mix position for *Phantom*, which has been running for more than 20 years and has crew and musicians and cast who have been on the show from the beginning. Mick was explaining that he did not believe in using a script ever when mixing. In England, it is common for the mixer to be in rehearsals with the cast for months mixing the rehearsals, which makes it easier to move into tech and not need a script. Mick believed that you have to feel the mix and if you are reading a script there is no way to feel it. I said I thought it was possible to achieve that while mixing with a script, which he adamantly disagreed with.

Recently he told me that he went back to hear one of his shows after several months and was unhappy with how the mix had drifted. I said maybe that wouldn't happen if there was some documentation of the mix. He disagreed and then explained that mixing was like playing music, and music always sounds better when musicians put down the music and just play from the heart. Jokingly I said I agreed with him, and to prove his point I was going to go into the pit and take the 20-year-old handwritten sheet music away from all of the musicians in the pit just to hear how they sounded.

When you work on a show, you will usually be given a copy of the script in a Word file from stage management. Sometimes all you are given is a hard copy. As a mixer, there is a lot of information you just don't need in a script, and the less text on the page, the easier it is to read. Most of the stage directions are worthless to a mixer. The mixer doesn't care that Steve crosses downstage and picks up a gun. So one of the first things to be done is to go through and delete any extraneous text or blank space. But wait—before you do that, you need to think about pagination.

The script will most likely have a footer with an automatic page number on every page. If you go through and chop up the script, your page numbers will not match the stage manager's script. This can make tech miserable. Just imagine trying to quickly find where they are starting from and the SM says page 58 and for you it is page 46. So, the first thing to do is to resolve the pagination issue. The only way to do this is to start on the last page in the script and put the page number in the body of the script. I usually put it on its own blank line at the top of the page aligned to the right. Work backwards until you get to the first page and then delete the footer page numbers. Now as you go through and alter the script you will end up having page numbers that are accurate but not necessarily on the top or bottom of the page. It is possible to have two or three page numbers on a single page if you end up cutting a lot.

After cutting this extra stuff, I tend to make a "style" for different aspects of the script. I will make a style for "Character Names" and one for "Scene Titles" and one for "Song Titles." Then I will go through the script and assign everything to the proper style. Once I do that I can easily select all items of a certain style and make it look like I want it to. I personally like my character names aligned left and larger than the dialogue. I also usually make them a color that pops out so it is easier for me to read, and I underline it. Once I have done that, I will go through and add DCA numbers for every character name. I will also highlight yellow things that I think will sneak up on me. If there is a scene with two people for a couple of pages and then suddenly someone else enters, I know I have a tendency to miss that entrance, so I will highlight that person's line. I will then start adding boxes with cue numbers and notes on the right side. My script will continue to grow through tech. Eventually I will hand it off for my sub to learn. I will show him how to quickly reformat things to his liking and let him make it his script.

VCAs/DCAs/Control Groups

In the olden days of analog mixing, we had something called VCAs on large format desks. A VCA is a voltage-controlled attenuator. This is a fader that you can assign multiple faders to and it works like a master fader. A VCA is different from a Sub-Group in that you cannot send the VCA signal anywhere. Instead, the VCA is controlling the level of the channel faders, but you still independently send the channel faders to Auxes or Mixes or Sub-Groups. A VCA is a crucial element to mixing. VCAs allow us to assign all of the men to one fader even though half of the men go to Group 1 and the other half go to Group 2. By using VCAs, we can create a logical and manageable area on the board to mix the show.

Once we moved into the world of digital consoles, the term VCA was no longer accurate. A VCA was actually controlling the amount of voltage allowed to the inputs in its group to adjust the volume of those inputs. The digital world is not dealing with voltage on the input level. When Yamaha released the PM1D in 2000, they introduced the DCA or digitally controlled attenuation. Other digital consoles have come along and introduced their version of the digital VCA and have given them other names. Control Groups is another popular digital interpretation of the analog VCA. Whatever term is used, the basic function is essentially the same. Don't be shocked if you hear an old-school guy like me talk about doing the VCA programming while working on a DiGiCo's Control Groups. For simplicity, I will be using the term DCA for the rest of this chapter, since most of the boards are digital now. You can adapt it to whatever style console you have. Whether you are using a Cadac with VCAs or a DM1000 with Master Control Groups, the ideas are the same.

A majority of mixing on Broadway is considered DCA mixing. We mix almost everything using the DCAs. If we leave the DCA section and grab something at the input level, it is sometimes called "board mixing." One reason for mixing on the DCAs is to make something complicated easier, but another main reason is to leave a majority of the board open for the design team to work. If you are mixing on the input faders, it will restrict where the designer can go to make adjustments. Most designers want the ability to make adjustments to the input levels that can be stored for a song. Think about a song with 30 people singing and a band of 24 playing. The designer will want the mixer sitting calmly in the middle of the board mixing something like two DCAs, one for men and one for women, and also one or more faders for the band. Then the designer can move around the board and solo different inputs and work on a balance of all the vocals and band. The designer can make adjustments to the inputs and then the mixer can save that snapshot so it can be recalled night after night. The mixer's job is simplified and repeatable. He puts the faders at −10 and mixes, but the inputs are all over the place to make it sound balanced.

In order to program DCAs for a show, you need to understand some basic philosophies involved. Consistency is very important in DCA programming. You want to make decisions about the programming that can be repeated throughout the show. You may decide that DCA 7 will be Men and DCA 8 will be Women. If that is the case, then you will want to repeat that every time you need a Men and Women fader. If you switch them in one cue, it is possible that you will bring up the wrong fader at the wrong time.

The first thing to think about when planning your DCAs is how you are going to treat your band. Will the band be one DCA or more DCAs? Will the band reverb return input be a separate DCA or will it be in the Band DCA? Different mixers like different things. There is no right or wrong here. Some mixers like breaking the band up into sections on the DCAs so they have some individual control from the DCA section. Typically, the band is on the right side of the DCA section, or the higher numbers. In this style you will see something like DCA12-Band, DCA11-Drums, DCA10-Solo. When used, the Solo DCA is a place reserved for instruments that may need to be pushed out for solos. When mixing on this style of programming you typically push all three DCAs as one and occasionally push or pull different DCAs to texture the mix. Other mixers prefer to keep the band simpler and just have a Band DCA. This can make mixing easier in some ways but more complicated when you need to push solo instruments out. Sometimes the way you do the band is dependent on how many DCAs you have. If you only have eight DCAs, then you probably don't want to waste three or four of them on the band, or you might not have enough to mix the rest of the show.

Once you have figured out the band, you can move on to the vocals. The general idea is to find an easy way to mix something complicated. Most of the time, mixers will put big block numbers above DCAs so when they look down they see numbers and they can just throw the numbers, as it is called. If you look at Figure 16.1 and 16.2, you will see DCA numbers by each character name. I gave those numbers for very specific reasons to make the scene easy to mix. When I mix that scene, I can look down at the board and mix the pattern. 1, 2, 3, 2, 3, 4, 3, 3&4.

Of course, while I mix that scene I have to put the band at –10 for a percussion sound, then take the band out and then preset it for the song. Because of the speed of this

2-BILLY
Excuse me, sir, where is Steve.

Band to -10.

7-CAPTAIN
Steve is in the brig.

2-BILLY
What did he do?

Band out.

7-CAPTAIN
He stole a woman's shoes.

6-Kanga
A woman's shoes.

7-CAPTAIN
Her shoes.

Band to -10.

6,7-Captain & Kanga
HE STOLE HER SHOES
THE MAN LIKES SHOES
WOMENS SHOES.

Figure 16.1 A script of a fake show the way I lay it out.

1- SAILORS
Whistle.

2-BILLY
Excuse me, sir, where is Steve.

Band to -10.

3-CAPTAIN
Steve is in the brig.

2-BILLY
What did he do?

Band out.

3-CAPTAIN
He stole a woman's shoes.

4-Kanga
A woman's shoes.

3-CAPTAIN
Her shoes.

Band to -10.

3,4-Captain & Kanga
HE STOLE HER SHOES
THE MAN LIKES SHOES
WOMENS SHOES.

Figure 16.2 A script of a fake show with DCA numbers.

show, I went with the band on one DCA because I couldn't manipulate three DCAs during this quick dialogue. We take the band out every chance we get, because 50 open mics do not do much to make your show sound good, and inevitably you are going to hear the musician bang something. I was the US Associate, or United States Sound Design Associate, on a tour and the mixer parked the band at −10 for the entire show. It sounded awful. There were 20-minute dialogue scenes with no music playing and yet he left all the band mics wide open. He actually went to the conductor and told him that the musicians needed to be quiet because they could be heard blowing their noses in the house. I suggested he take the band fader out and he refused. He didn't last long.

As you plot out your DCAs, you will want to take your script, sit down, pretend to mix the show, and figure out how it will feel to mix. As you do this, it is good to mark the script as you go and build a table that defines each cue (Figure 16.3). Typically, I will have the script and ShowBuilder open to document the DCA scenes. Once I have marked up my script and created this table, I am ready to program the board.

When you program your DCAs, there are some good tricks that can be employed. First, you want to make it easy and stress-free to recall cues. Unless it is absolutely necessary, you want to place DCA cues in long speeches, and if possible, do not make them crucial to be fired on an exact moment. This is not always possible, but it keeps the stress level down when you are not panicked about hitting the cue. If you do have to fire a cue in the middle of some quick

Figure 16.3 The DCA scenes for a show.

VCA Library	1	2	3	4	5	6	7	8	9	10	11	12
2.0	Reno	Billy	Whilney	Fred	Reporter	Purser	Captain	Sailors	Vverb		Band	
3.0	Reno	Luke & John	Minister	Photographer	Reporter	Purser	Captain	Angels	Vverb		Band	
4.0	Purser	Billy	Whilney	Photographer	Reporter	Mrs. Harcount	Hope	Evelyn	Vverb		Band	
5.0	Purser	Billy	Minister	Luke & John	FBI 1	FBI 2	Erma	Moon	Vverb		Band	
6.0	Purser	Billy	Whilney	John	Luke	Captain	Erma	Moon	Sailor		Band	
7.0	Purser	Billy	Leads	Men	Women				Vverb		Band	
8.0	Purser	Billy	Reno	Hope	Evelyn				Vverb		Band	
9.0		Billy	Reno	Hope					Vverb		Band	

dialogue, it is helpful to have a couple of DCAs carry over from one scene to the next. Let's say you have two people talking quickly and you need to take a cue to set up for the next people talking. If you have the first two people carry over into the next cue, then you can take the cue any time in their conversation without it being an urgent cue. It can also be a good idea to have people in a cue who aren't really needed until the next cue; that way, if you throw the cue late you will still have the first couple of people you need.

Both of these tricks allow you to extend the window for when you need to throw a cue. When you use one of these tricks, you can give yourself a visual cue on the board. When you name your DCAs and you have some DCAs that aren't needed in the cue, a good idea is to label those DCAs in lowercase. Then when you recall the cue that does need those DCAs, you can change the label to using uppercase letters. This will give you a visual cue that you have recalled the cue and you are in the right scene. Also, you can add a note in the DCA label for the level of that fader. We do this a lot for the band DCAs or other faders that need to be preset. By putting a –5 under the name, you give yourself guideposts to follow while you mix just by looking at the DCAs.

Console Programming

To start programming your console, you need to understand the rule of the 0 or Unity Gain. If you walk up to almost any board on Broadway, you will find that all the inputs are set to 0 and turned off. That is where we typically start every cue or snapshot. This allows us to assign a VCA, turn the fader on, and be done with that scene. It works on an aesthetic level as well. When you are mixing and look down at your board, and everything is a 0 or Unity Gain except for a couple of faders, then you can easily see what the boosts or cuts are. If you are using a board without motorized faders, you want to park the inputs somewhere that is easy to re-establish. If you are using a board with motorized faders, you don't really want it to have to move faders up and down all night long. Sometimes that is noisy, and it will eventually wear out the faders.

It is a good plan to create a cue that is a base point for your show, or a cue that can be recalled to wipe the console back to a neutral state. This typically involves having all the inputs at 0 or Unity Gain. A lot of people call this the base cue, and it can take a long time to create. Take your time and create a very accurate base cue. In this cue, you will want to name every input and output. You will want to route all of your inputs to Groups and set the Aux mixes. You will also want to do your Mix to Matrix section. We are Matrix obsessed on Broadway. The more Matrix, the better. Typically, we send our Inputs to Groups and our Groups to the Matrix and the Matrix to the speakers. There are different ways to do the Groups, but the basic idea is to have Groups for the Vocals and Groups for the Band and Groups for the Sound Effects. That allows us to create different mixes quickly to different Matrices, depending on what is needed in that zone. One style of grouping is 1-Men, 2-Women, 3-Band Left, 4-Band Right, 5-Drums Left, 6-Drums Right, 7-SFX Left, 8-SFX Right, 9-Gods. This gives some good options. We can EQ the men and women differently, and by breaking the drums out into their own groups, we can balance the drums in different locations. Another style of grouping is 1-Head, 2-Ears, 3-Band Left, 4-Band Right, 5-SFX Left, 6-SFX Right, 7-Gods. The Head Group is for any lavs that will be worn on the head. The Ears Group is for any mic that will be worn on the ear. This allows you to quickly find a decent EQ for everyone based on mic position, and then individuals can be fine-tuned.

Once you have labeled and assigned everything, you have to develop a strategy for "Recalling." When you are teching a show, there are times when you want everything to recall and times when you want nothing to recall. All digital boards have the ability to save and recall scenes. Most digital boards have the capability to decide whether you want certain parameters to recall in only some scenes or in all scenes. There are times when you may want an EQ to recall to the stored value and other times when you don't want it to recall. There are times when you want every aspect of the board to recall and other times when you want only some aspects to recall. Developing your recall strategy is probably the most important part of programming your board. If you do it wrong, you are going to be constantly trying to figure out why you lost settings and the designer is bound to get very frustrated after he finishes quiet time and finds out that you accidentally erased all of his delay settings. There is definitely no right or wrong here. Every console works differently and every designer wants something different. What is important is for the mixer to understand the recall strategy and to make sure nothing accidentally gets lost.

My personal strategy is to start with very little recalling and increase what the board recalls as we move through tech until the show opens and is frozen, and then basically recall everything all the time. When I start, I make sure nothing in the output section recalls. This works for most designers because they don't really vary the EQ and Delay of the mains that often. There are always exceptions, but it is generally better not to recall the output section at first. The same goes for the band inputs. It is generally a good idea to keep the band inputs static until a basic mix has been found. I usually also safe out the EQ, Compressor, and Gain. The term *Safe* is used by some digital desks, such as Yamaha. Other desks use different terminology, but the concept is the same. It means to tell a certain parameter not to recall the stored parameter when a scene is recalled. There are typically two types of Safe: Scene Safe and Global Safe. Global Safe means it is Safed in all cues. Scene Safe stores in a scene and tells the parameter whether it is allowed to recall in that cue or not. When I start tech, basically the On/Off and Fader level of my RF inputs recall with every new scene or preset and that is about it. As the system begins to develop, I will start recalling other things. Once we open, I will generally recall every scene one at a time and store it. Then I can turn off any Recall Safes and have the board constantly recall everything. If I ever need to make a change during the run, I can easily safe out whatever I need to change.

Once you have your strategy, it is time to program your DCAs. If you have laid the DCA scenes out in a table and marked up your script and built your base cue, this should be very quick and easy. My method is to recall the base cue and then save it as whatever the cue number in the show is going to be. Then I label the DCAs, program the inputs to the DCAs, and turn those inputs on. Then I store the cue, recall the base, and repeat until I am done. It is very important that you assign a DCA and then turn it on.

If you need to take something out of a DCA, you turn it off first and then out of the DCA. If you do not follow this order, you will open an input up wide, which could cause raging feedback.

As you program this, you will find questions you can't answer, like who will be saying the lines for Sailor 3 and 4. Hopefully the stage manager will give you a breakdown of what ensemble members will be saying or singing certain lines, but if not, you should make a list of questions to ask the SM as you program. It is very easy to go back later and assign and turn on the appropriate people. Once you have your DCAs programmed, you should be ready for cast onstage. It is not always possible to be completely programmed before the cast starts onstage, but it is a good idea to have as much done as you can. The more you have programmed, the less stress you will have.

Playback Programming

No matter what playback system you use, the most important question for the mixer is how it will be triggered. Since the programming will basically be done for you, the only real issue is how you will fire the sound cues. Do you want one button that fires the scene on the board and the sound cues, or do you want a separate button, one for sound cues and one for board cues? There is an advantage to both. If you have a separate button for sound cues, then you are less likely to accidentally fire a sound cue. If you use only one button for both, then you simplify the mix, especially if there are lots of sound cues.

It is very common for the design team to program the playback system, and the mixer will have very little to do with it. According to union rules, the design team is not supposed to touch equipment, but that is a very challenging concept in sound. If you look at the lighting model, you can see what we are supposed to be. In lighting there is a lighting programmer. The lighting designer talks to the programmer and explains how he wants something to look and the programmer programs it for him. The LD will tell the programmer to bump up certain lights by a certain percent or change some other parameter, and they will chat with each other on headset constantly.

Could you imagine trying to mix if the sound designer put you on headset and talked to you the whole time, asking you to give a 2dB cut at 160 with a Q of 1.2 on Steve? It would be impossible. Or think about how many times you've heard the LD say he is cueing ahead or behind. Could you imagine if we were programming ahead and had live mics open of people offstage so we could fix their EQ for an upcoming scene? There is an understanding that what we do is different and requires more hands on from our designers, but the union is entitled to ask the production to hire a "programmer" to deal with this. If you have a programmer on contract, which means a union person is being paid to program the equipment just like in lighting, then people have less of a problem with sound designers touching gear.

The reality of what normally happens, though, is that the playback system and console are programmed by the designer, mixer, or assistant. It depends on how the designer works. There are shows that have a person dedicated to firing sound cues, which was the case on *A Christmas Carol*, *Fela*, and *Bombay Dreams* when I mixed those shows. More often, all cues are fired by the mixer.

17

A2'S JOB (MIC'ING TECHNIQUES)

As sound people, we are required to know a plethora of obscure and seemingly unrelated information. We are also ridiculed for our endless fount of useless knowledge. Two sound people can hold a lengthy conversation using only numbers and letters, which has caused my wife to boycott dinners with more than one sound person in attendance. We are required to know basic electronics and microphone techniques. We must know how to mix and what TV stations are broadcasting. We need to know that midi can only travel 50' without an extender and how humidity and temperature affect the sound we hear. We also must know how to properly attach a lavaliere to an actor and how to prevent sweat-outs. We should know the name of a good medical supply company as well as a place to get cheap defective condoms. We should be able to hold a meaningful conversation with the nurse in the delivery room about Tegaderm, although my wife doesn't agree with this. It really is a profusion of minutiae. We also must possess the strength to make everyone happy, the wisdom to know what to say when they are unhappy, and the patience to know we will never make them happy. And, of course, we have to know intercom and video.

Part of that minutiae includes knowing what equipment you need in your RF toolkit and the secrets of dealing with wireless mics. I talked to one of the most experienced deck sound people on Broadway to get the lowdown. Mary McGregor has worked as an A2 or deck sound on dozens of shows. She currently works at the Metropolitan

Opera and has recently worked on *Book of Mormon,
Spring Awakening, Chicago, In the Heights, Spamalot,* and
dozens of other shows. McGregor explains how to rig and
place mics:

> *The optimal placement of a mic on an actor seems to be the
> center of the forehead. The first questions to be asked and
> answered are: Is there hair or no hair? Is there a wig or no
> wig? Is there a hat or no hat? These things help determine
> the actual placement and rigging parameters. Tailoring a
> mic to an actor, the next step, moves into the "craftsy" area of
> being an A2.*

She explains the arts and crafts:

> *The basic three toupee clip rig with the mic head at the
> center of the forehead will work for a healthy head of hair.
> If we are using a beige mic, it is necessary to paint the cord
> to match the actor's hair. I really like the Copic art markers
> best for this. When they fade due to sweat and hair products,
> the color isn't horrible. Pantone and Letraset art markers
> are also good. Do not use a black Sharpie EVER. It will
> always go blue. Color the mic the length of the actor's head
> from forehead to nape. Add 2" of Craft wire in a color that
> matches the hair color at the element end. This helps to
> focus the mic toward the actor's mouth. String three toupee
> clips onto the cord. The clips should be prepped with 1/8"
> elastic tied through the holes. The clips should come close to
> actor's hair color. I also paint the very head of the mic with
> black/brown/beige nail color. This way the color won't sweat
> off onto the actor's forehead.*

A bald or balding actor is a challenge. For this, McGregor
suggests an ear clip:

> *Over the ear is the next best placement. I have placed
> mics on the temple using the standard 3-clip rig. With a
> DPA, I like the clear plastic ear clip. With a MKE2, I like
> the heavier metal clip. I try to get the head of the mic on
> the actor's cheekbone away from the sideburn. Craft wire
> on the mic for the distance from the ear to the cheekbone
> helps hold the mic close to the face. Using a Hellerman tool,
> attach the mic cord to the ear clip with 2 or 3 Hellerman*

*Sleeves. If the actor has hair on the back of his neck,
you probably need a small toupee clip after the ear clip.
Otherwise, plan on taping the mic behind the actor's ear
and at the neck.*

She explains how wigs make life easier:

*If the mic is worn with a wig, very little needs to be done.
I have had quite a bit of success with putting them under the
lace front of wigs. The head and maybe 2" can be painted to
match the hair color or skin tone. I have also had the head
peek out from a hard-front wig. Sometimes, depending on
movement, it may be necessary to go to a hi-boost cap on a
DPA, because the curved screen on the soft-boost cap scrapes
against the lace.*

McGregor has found an inventive way of painting mics
temporarily:

*Lately, I have been dealing with mics in a "rep" situation;
I have started carefully covering the mic cord, where it gets
painted, with paper adhesive tape and painting the tape.
Then, when the show is finished, I can remove the paper
tape and have a plain mic to use for the next show. The
paper can last two to three shows but will need to be redone
for more shows.*

McGregor explains the delicate art of how to stick the mic
to the actor's skin:

*Generally I put one small piece of tape on the actor's neck.
This is mostly to strain relief the last toupee clip so it won't
be pulling on the actor's nape hair. It's also to counter-act
"loopage." That is, when the cord "loops" out of the actor's
collar and can be seen by the audience. Three kinds of tape
are in general usage: Transpore, Blenderm and Tegaderm.
I have also used paper tape, fabric tape and band-aids.
Whatever the actor is comfortable with, and doesn't irritate
his/her skin is what I go with. The last choice is Tegaderm,
because it is so expensive. It does, however, come in large
sheets for naked or plunging back taping. If the actor is
sweaty, you can use alcohol pads or skin prep pads to wipe
the neck or ear area before taping. This will give you a drier
surface for the tape.*

And what can you do about those pesky sweat-outs? McGregor explains,

Sweat control is another actor-specific situation. Moleskin, available at drug stores near the shoelaces, is the easiest to find. Elastoplast is better because it doesn't get furry, but it is harder to find. I take a small piece of the Elastoplast the length of the mic head's diameter and wrap it around the cap not covering the windscreen. These wraps need to be changed regularly as they become messy and stop absorbing. If the actor is a heavy sweater and the mic belt is next to the skin or a really wet undershirt, the pack might need to be bagged. The cheapest are sized plastic bags taped closed. A cotton puff or two at the top around the connectors can be added before you tape the bag close. Sheathes or condoms can also be used.

I don't like long-term condom use because the talc can gunk up the battery compartment and the off/on switch on the transmitter. Sheathes, ordered from a medical supply company, don't have the talc.

But your job is not done when the curtain comes down as McGregor explains,

Maintaining the microphones after the show has been set is also very important. I like to use non-acetone spirit gum remover on the cords to remove tape goo. Krylon makes one and so does Ben Nye. Most hair products can be removed with soapy water. Painting always needs to be touched up, and the elastic on the toupee clips should be checked often. The windscreens can also get clogged with sweat. I like to use an ultrasonic jewelry cleaner for the caps. A $60 one is usually just fine. The battery operated ones aren't powerful enough, and the ones with digital controls and heat are too much. Mr. Clean, a low suds anti-bacterial cleaner, is fine, but there are others on the market.

For hygienic reasons, everyone should have his or her own mic. Budget constraints may limit this possibility.

McGregor explains the differences she has found among designers and actors:

I find that some designers care more about hiding the mics than others. Some actors care more about the

comfort of the pack placement than others, which can have more to do with the costume than anything. The bottom line is to maintain the sound quality. The main job of an A2, in my mind, is maintaining the consistency of that sound. Once the EQ is set for a particular actor with a particular placement, every show thereafter should be the same, even if there is just one rehearsal and one show. Live theatre is a fluid thing and no two shows are ever the same performance wise. The mic placement on the actors at least takes one more variable out of the equation so the mixer doesn't have to chase EQ and attenuation all night.

Items Mary McGregor recommends in your RF Toolkit (Figures 17.1 and 17.2) include:

Masque Mic Check Box: MTB-50
Battery checker
Hellerman tool
Hellerman sleeves 15 mm & 20 mm beige and black
Craft wire: black, eggshell, white and gray 20 ga and 22 ga
Wire cutter/pliers
1/8" elastic black and white
Toupee clips in black, brown, and beige in small, medium, and large
Nail polish in black and shades of brown; also clear
Selection of ear clips
Alcohol pads
Skin prep pads
Art markers: Copic, Pantone, or Letraset in "hair colors"
Tape: Transpore, Blenderm, and Tegaderm
Krylon MME-mic cleaner
Ultrasonic cleaner
Sharpie, for labeling only
Cotton puffs
Safety pins
Scissors
Rat-tail comb
Selection of tie clips, barrettes, and bobby pins
Plastic bags and/or condoms

Figure 17.1 Some of the supplies in Mary McGregor's RF Kit. The following list shows the definitions of the tag items in the picture.

A—Hellerman tool with black/beige sleeves
B—Black nail polish and craft wire
C—Elastoplast in beige
D—Coloring pens
E—Tegaderm, Transpore, and Blenderm
F—Mic testing box MTB-01
G—Sheathes and plastic bags for transmitters

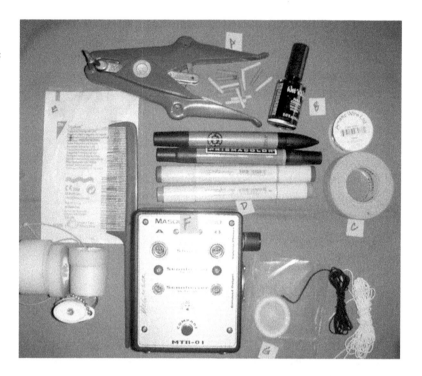

Figure 17.2 Some rigged mics prepared by Mary McGregor. The following list shows the definitions of the tag mics in the picture.

A—Elastic headband
B—Double rig with Hellerman sleeves and Elastoplast on the heads
C—Mic painted black with craft wire on the front and three small toupee clips
D—Ear clip with Hellerman sleeves, craft wire on the front and one small beige toupee clip
E—Brown painted double rig with craft wire on the front and three small brown clips.

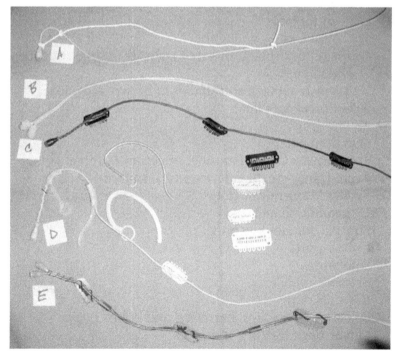

18

TROUBLESHOOTING

I was doing a show with Brian Ronan in a medium-sized theater. It was a small-scale musical, the kind I have done a thousand times before. Tech had gone pretty smoothly. So smoothly that I couldn't even remember going through tech. But things did not go according to plan on one preview. The show started out just fine. I was mixing away and having a good time. There were tons of sound cues on the show, which is not that common for a musical. Normally musicals are some doorbells and thunder cues unless you are doing something like *Shrek* or *Young Frankenstein*. We were running timecode and midi to fire the sound effects. Sometimes the cues would get triggered by lights and sometimes we would trigger a timecode sequence that fired off cues. Nothing fancy. Just some automation. But on this particular preview that system all started to fail while I was mixing. Brian was at my side as I struggled to figure out what was going on and keeping the audience from figuring out what was happening. I had the ability to fire everything myself, but moments were approaching that were going to be impossible for any one or two human beings could do flawlessly on the first pass. Luckily Brian is a top-notch mixer in his own right so if I told him I needed to crawl behind the racks to fix the problem, he could hold down the fort on the faders. And that's just what happened. In the middle of Act 1 I scampered under the racks looking for a cable that I suspected was the problem. If I was right, I would reseat the cable and all would be fine and we would not be stopping the show.

And then I woke up. For a few minutes of hazy consciousness, I laid in bed continuing in my head to think through the signal flow of the midi trying to pinpoint the problem. Finally, I shook my head and reminded myself that it was a dream. There was no problem. There was no show. In fact, I hadn't mixed a show for Brian in well over a year. I did email him that morning to tell him of our near-disaster experience and to assure him that I knew what the problem was and that I could have fixed it if I hadn't woken up.

To me, this is what it is like to be a mixer. The stuff of nightmares for me as a mixer is not missing a pickup; it's troubleshooting on the fly while not missing a pickup. My friend Jordan equates mixing with boxing and I find that to be a good analogy, but I equate it more to being a World War I fighter pilot. Here I am flying this plane. A well-designed and well-built plane, but a plane with weak points, like lavaliere mics. And as I am flying, parts of my plane start falling off. A good WWI pilot would be able to land that plane with just one wing and no motor left. That's how I see mixing. You have to fly that plane no matter what breaks and you have to land it safely at all costs. We do musicals with excellent equipment now. Even on the low end, the Berhringer M32 is a fantastic and solid piece of equipment. Our systems have good speakers. These are solid systems. But at the end of the day all of our systems can be defeated by one bead of sweat. One! Or one broken mic. A skill that is essential for a theatrical sound mixer is troubleshooting. Without it, you will fail in this business. A mixer has to have a strong understanding of the basic methodology for troubleshooting. Also, the ability to predict and plan for failures and have a plan for rescuing the show from failure.

This chapter is going to break down the most fundamental rules of troubleshooting for different areas of a sound system, and at the end I will throw some fun scenarios at you to test your skills. I will break the sound system down into six categories: Inputs, Outputs, Intercom, Video, RF, and Midi. There is one other category that is important, but not exactly sound related and too varied for this chapter, which is networking. In modern sound systems, there is a ton of networking, but I will leave that to a book like *Networking for Dummies* or something like that.

Inputs

The first and most basic category is inputs. Without input, we are nothing. Of course, there are times when we get blamed for what that input is, but mostly we do not create the input; we just amplify it. I had a director last night give a note to his cast that they were bringing in a diction coach and that the mics sounded great but made the poor diction more obvious. He's right. We can't fix the fact that an actor says a "t" like a "d," but we sure can make it louder. I have also known sound people that call inputs "guzintas," as in "goes-into." I will not ever refer to inputs this way, nor will I ever refer to outputs as "guzoutas." If we ever have the good fortune to work together and I could make one request, it would simply be please do not ever use those words in front of me. I do not know why, but it's like fingernails on a chalkboard. So let's get cracking on inputs.

The first rule of troubleshooting anything is to start at the beginning. It's a very good place to start. (Sorry about that, but I am in Yakima, Washington right now, designing the *Sound of Music* tour.) Maria is correct. You start as close to the beginning of the chain as you can. If you are checking your band mics and the snare drum is dead, then you start looking for the problem at the console and work your way toward the snare drum.

But don't let yourself get bogged down with the first problem you run into. Bang through your whole drumkit. Make note of the problems and then go back and fix the problems. If you do that, you might find that no drum mics are working because the mult hadn't been plugged in or the wrong mult was plugged in. Once you have verified which inputs in the band are either not working at all or not working properly (such as buzzing), then you can go back to troubleshoot the problem mics.

First question—what kind of mic is it? Is it a condenser? If so, then double check that you have phantom power turned on. Next, plug a phantom power device into the input hole on the console. In the olden days of analog that would mean you go around to the back of the rack and plug something in to make sure the input hole is actually working. In modern times we have remote racks onstage or in the pit that are the actual input holes. Step two for me is always go straight to the rack. If it works at the rack or console input, then you have verified the problem is somewhere in the cabling or is the mic itself. Next you plug the mic back into the console and check at the next breakpoint. If this is a snake or a mult, then you plug your phantom-powered device into the line on the snake or mult. If it doesn't work there, then you have verified that you have a bad mult line. Move to a spare at the console and check the snake on the new hole. Then plug the snare mic back in and see if the problem is solved. If not then move to the next breakpoint, which is most likely the mic itself. Plug your device in instead of the mic. If that works, then you have found a bad mic. Yes. Mics die. Replace the mic and move to the next problem.

I know this probably seems straightforward and logical (because it is), but it is important for you to burn these steps into your head. If you start troubleshooting at the end of the chain and work your way back to the start, you will get very frustrated when you've wasted 20 minutes only to find that phantom power wasn't on. As far as a phantom-powered device, you can use a condenser microphone as a tester or you can get something like a Noise Plug (www.tangible-techno logy.com/gtc/Noiseplug.html).

I am a huge fan of a Noise Plug (Figure 18.1) because you can carry it in your pocket and it isn't very expensive and doesn't require batteries. You simply plug it into an input and a red light comes on if the Noise Plug is receiving phantom power. Plus, it sends pink noise down the line so you can verify at the console that the signal is getting to you and sounds good. It is one of the many small tools that I always have with me.

Another good input tester to have is a Rat Sniffer (www.rat soundsales.com/p/soundtools-xlr-snifsend.html).

The Rat Sniffer (Figure 18.2) is really good at helping test for bad cables once you have run out the cables. It is also a fairly cheap tool to throw in your workbox. The Rat Sniffer is a two part XLR barrel device. You plug the part with one light into one end of a cable and you plug the part with three lights into the other end of a cable. The Rat Sniffer then sends voltage down all three pins of an XLR and the "A," "B," and "C" lights will turn on to show that all three pins on the cable are good.

Figure 18.1 Noise Plug.

It should go without saying that you need a good cable tester. Hopefully your cable was tested before it was run out, but that is not always the case. If everything else seems to be working, then it is time to pull out a cable tester and see if the cable is the problem. My favorite cable tester is the Whirlwind MCT-7 (Figure 18.3; http://whirlwindusa.com/catalog/black-boxes-effects-and-dis/testers/mct-7-cable-tester)

Figure 18.2 Rat Sniffer.

This cable tester tests XLR, 1/8" TRS, 1/8" TS, ¼" TRS, ¼" TS, midi, RCA, BNC, NL2, and NL4. That's a whole lot of tester in one little box. And it tests each pin of each type of cable for continuity and mistaken cross-wiring. Plus, you can test adaptors. You can test an XLRM to ¼" TRS. Or even a midi to XLRF. It is very common to convert midi to XLR because it is easier to get a 25' XLR than it is to get a 25' midi cable, and you only need three of the five pins in the midi cable in order for it to work. The other two pins are for future expansion. I am not sure what they will expand to or when. I consider those two pins to be like Dipping Dots. They are the pins of the future.

Figure 18.3 Whirlwind MCT-7.

There is another tester box that is becoming a standard with a lot of sound people. It is a more expensive box, but it sure does make a big impression for a box of its size (that is a musical reference to a song by the Old 97's called "Big Brown Eyes"). The tester is the CTP Systems dBbox2 (Figure 18.4; www.ctpsystems.co.uk/dbbox2.html).

The dBbox2 is a cable tester for midi, XLR, and BNC. It also takes analog or digital signal in using XLR or BNC and routes that signal to an internal speaker. It can also send signal out either analog or digital using XLR or AES. It can also read incoming midi messages and produce midi messages to send. There are other functions as well. This box is a monster and has become my single favorite tool in my bag. If I were to be stranded on a desert island with a Broadway musical size sound system and I could only have one tool with me to make the system all work, this would be the tool I would choose.

Figure 18.4 CTP Systems dBbox2.

Outputs

The next category is outputs. When I test a sound system, I usually check my inputs first, once I have the inputs working and all this pent-up signal that needs to get out; then it is time to move on to the outputs. The basic principle is the same. Start at the beginning of the signal chain. With all troubleshooting, our goal is to eliminate as many variables as possible. We test one variable at a time until we have added all variables back in and have found the offending item. With outputs, we start at the console and work toward the speaker.

In order to check outputs, we want a consistent and solid source of signal. Usually that means pink noise. I almost exclusively check outputs with pink noise. I don't like the sound of it, but it is the most practical choice for testing. I know lots of people using music to test outputs but that does not work for me. The reason is simple. Pink noise is consistent. Pink noise always sounds the same. Pink noise is signal made up of equal energy in all octaves of frequency, which means, unlike white noise, it follows the logarithmic nature of human hearing. If I turn pink noise on and send it to a speaker I can tell if the highs are not there. If I, god forbid, turn on some Steely Dan and send it out to a speaker, I can't tell if there is a problem with the speaker or with the crappy song I am listening to. If I bounce from one speaker to another with pink noise, I can tell if there is a difference between speakers fairly easily. If I do the same with "Babylon Sister," as I heard one designer do for a solid eight hours . . . multiple times, I can't tell if there is a problem with the speaker or if I am just in a part of the song with less high-end information. Moral of the story is, listen to music for fun. Test systems with pink noise like a professional.

Outputs can be trickier to test than inputs. That is why I usually do the inputs first. I like to feel a sense of accomplishment. Oh, and one bit of advice with testing. This is a very valuable piece of advice that will make your life much better. Never end your day by testing something unless you absolutely have to. After a long day of running cable and setting up speakers and mics, don't decide to spend the last hour checking stuff if you have time the next day. The reason is simple. It's like a Murphy's Law of sound. As sure as I am drinking this decaf cappuccino, I guarantee you that in that last hour you will find dozens of problems that need to be sussed out and you will be out of time to do any trouble-shooting. Instead of leaving work feeling good about the day and ready for some sleep, you will feel anxious about all the problems and you will troubleshoot in your dreams. No one wants that. Plus, you are more tired at the end of a call and more prone to making simple mistakes. It is always better to start testing stuff when you are fresh in the morning and more clear-headed to tackle the mundane task of practical logistic troubleshooting.

One reason outputs are trickier is that it is not always a straight shot to a speaker the way it is to a mic. The output of the console may have an EQ inserted on it. The output then probably goes to some processing unit. It might get split into two or three signals for lows, mids, and highs. It then could hit multiple channels on an amp or loop through from another amp. Then out of the amp to recombine to a single NL8 cable, then to a box, and then to the speaker. It is a much more complicated signal path. But just remember, the principles are the same. The first thing you want to do

Figure 18.5 Vizear SoundPlug.

is set up an input channel for pink noise. You can either use the Noise Plug into an input or you can look for an internal pink noise generator on the console. If I am using a digital console (which is always the case nowadays), a lot of times I prefer to send the pink to a Direct Out so that I am avoiding all pitfalls on the console and just send-ing directly out the output socket. If no signal comes out the speaker I expect, then the first place to start is by grabbing one of three tools. The first tool is a cheap tool that I love to keep around because it is small and doesn't require batteries. It is the Vizear SoundPlug (Figure 18.5; www.vizear.com/SoundPlug.htm).

This is a simple XLRF barrel with a piezo speaker built in. If you plug this into the output socket of your console and stick the rubber end into your ear, then you will hear signal if the output socket is working. Now, this is a simple tool that is only good for basic proof that signal is coming out. It is not meant for any kind of critical listening. The next step up from this is a tool from Whirlwind called the QBox (Figure 18.6; http://

Figure 18.6 QBox.

whirlwindusa.com/catalog/black-boxes-effects-and-dis/testers/qbox).

It is the same idea with the QBox. You plug an XLR from the output socket of the board into the QBox and turn the volume up on the internal speaker. This is a bit more appropriate for critical listening. You will be able to tell if the signal is good and strong. You will also be able to tell if you have a buzz or hum. You can also use headphones to check the signal. Alternatively, you can send a tone from the QBox directly to a device. For example, you can send a tone directly from the QBox to an amplifier and test the feed from the amplifier to the speaker.

You can also use the dBbox2 the same as a QBox. Once you have confirmed the signal is getting from the console to the processor and then to the amp, then you have to move to the other side of the amp if you still don't have signal. In order to do that, you can use any speaker and plug it into the amp. If you have no output to the speaker, then the problem is the amp. If you do have signal to the test speaker, then the problem is the cable to the speaker. An easy way to check cable to a speaker at each breakpoint is with a Whirlwind Cab Driver (Figure 18.7; http://whirlwindusa.com/catalog/black-boxes-effects-and-dis/testers/cab-driver).

The Cab Driver is an amazing little box to have in your kit. You can use it to drive signal to an XLR output or to speakers on NL4 or NL8. You can even select which pair on the NL4 or NL8 to send signal to. By using the cab driver, you can start testing a speaker at the speaker and work backwards to the amp until you find the bad cable. It is also very helpful when loading in a system and testing your speaker cable runs before your amps are in place and your system is fully setup.

A tool you will want for testing your speakers once you have verified they work is a Galaxy Audio Cricket (Figure 18.8; www.galaxyaudio.com/products/cpts).

The Galaxy Audio Cricket tests for polarity issues. With checking a passive speaker, you have to be very careful in wiring the speaker. If you wire the amp opposite of how the speaker is wired, then the speaker will be polarity reversed, which basically means that when the driver should be pushing out it is actually pulling in and vice versa. If two speakers are interacting with each other and they are polarity correct,

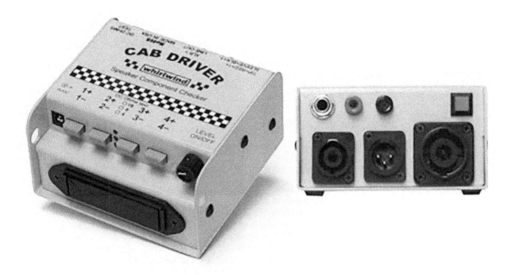

Figure 18.7 Whirlwind Cab Driver.

then the speakers will sum together and get louder. If one speaker is polarity reversed, then the two speakers will cancel each other out and get quieter and sound very unpleasant. One of my least favorite feelings in the world is hearing out of phase speakers.

The Cricket is made up of two boxes: a Sender and a Receiver. The way the Cricket works is you plug the Sender into the input of an amp or a channel on your console. Once you turn up the amp or the console channel, you will hear a popping sound that happens about every two seconds. You then take the Receiver and hold it close to the speaker you are testing. There is a mic built into the receiver unit. The Receiver picks up the popping sound and measures whether the polarity of the pop is correct. I typically check this from an input channel of the console so that I am checking the polarity of my entire signal chain. If there is a polarity issue then I will check the speaker from the amp so that I can eliminate all variables except the speaker cable.

You have to be careful when checking speaker polarity. This is not something you do unless you know what the speaker is supposed to do. A simple full-range speaker should always pop positive, but a bi-amped or tri-amped

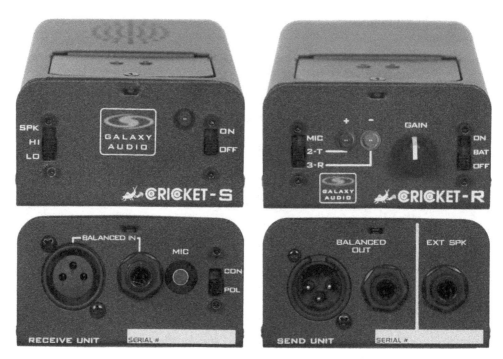

Figure 18.8 Galaxy Audio Cricket.

speaker can be very tricky. You need to research what the manufacturer says the speaker should do. Some tri-amped speakers are polarity-positive for the high and low driver and polarity-negative for the mid-range driver. People much smarter than you and I figured out what the speaker is supposed to do, so I recommend not going against what the manufacturer recommends for a speaker. If you are popping a bi-amped speaker that has a positive high and a negative low, then you could get different readings for that speaker depending on where you are holding the Receiver. In a bi-amped or tri-amped speaker, it is best to pop one driver at a time and verify that it matches the manufacturer specs.

You also have to be very careful in popping a powered speaker. Again, it can be deceiving unless you know what the speaker is supposed to do. The problem with powered speakers is that all of the processing is built into the speaker, so it is very difficult to test one driver at a time; if different drivers have different polarity, then you will get fluctuating readings depending on where you are holding the Cricket Receiver. A bonus to most powered speakers is that the only place a powered speaker can be polarity reversed in with the XLR. As long as your XLR cables are not cross-wired, then your speakers will not be polarity reversed. You can use the Cricket to test for this as well. Instead of using the mic to test for polarity, you can plug the XLR into the Cricket Receiver instead of the speaker. The Cricket will light up to show if the signal from the console through the XLR is positive or negative.

Midi

Most modern sound systems use midi somewhere in the system. I use midi to change reverb patches on reverb units. I use midi to fire sound effects. I use closed-contact buttons that the conductor pushes that converts to midi to fire click tracks. I use midi show control to allow lighting cues to fire sound effects. I use timecode over midi to have the light board fire light cues in time with beats in a song. I use midi a lot. Most people use midi a lot. Midi is crucial to the running of most shows, and yet a lot of sound people I know have no way to test their midi other than to test the midi cables. But knowing that the cable and all five pins are working, including the two unused pins, is just not enough.

There are times when testing midi that you will want to capture the midi message. If you have a reverb unit that is not firing when midi is received, but you know that the cable is good and that the midi receive light is blinking on the unit, then you are at a point where you need to capture the midi message to see what is being sent and received. You might find that midi program changes, which are the most common method of making changes using midi, is 1 to 127 on the sender but 0 to 126 on the receiver, which would cause the equipment to not respond correctly because the message you are sending is one number off from what the receiver is expecting. Or you might find that you have your sender set to send on channel one but the receive unit is set to receive on channel four, in which case nothing will happen with the receive unit.

Figure 18.9 Masque Midi Snoop.

In order to test the midi message, you are going to need a tool with the ability to send and receive midi messages and display the messages. If we first look at the ability to capture a message, there are several tools available. First is the Masque Midi Snoop (Figure 18.9; www.masquesound.com/product/masque-sound-midi-snoop-ms-40/).

This is a nice simple box to put in your toolbox. You just plug a midi cable in and it will tell you what kind of message you are receiving and on what channel. It will also tell you if it sees an MSC (Midi Show Control) message or a MMC (Midi Machine Control) message. It will also show if it is getting Timecode and will blink to show the code tempo. This is a great tool for a quick midi test. It does not go deep enough for some problems because it can't tell you the details of any message.

If you need to see details about the message you are receiving, then you have to move to another tool. You can plug the midi into a computer and use software like Midi-Ox (www.midiox.com/) to capture the message and translate it, but it's not all that logical to have to carry your computer and a USB midi device everywhere you need to test midi. Sometimes you need to test a connection in the back of a rack and using a computer is not ideal. You can also use the multi-versatile and awesome dBbox2, which will capture a midi message and tell you almost everything you need to know about it.

There are times, though, when a program like Midi-Ox is your only option. I was trying to get an Eon light board to trigger sound cues on a show and everything was testing fine. I was seeing blinky lights everywhere I expected to see blinky lights. I confirmed I was getting an MSC message from the light board to the sound console. Everything should have been fine. Eventually I had to capture the MSC message in Midi-Ox, which it displayed properly as a hexadecimal message. I then had to translate the hex message to figure out why the cues weren't firing and found out that this particular light board programmer was adding a ".0" to all cues. MSC is very precise and has to be an exact match. He was telling me he was firing cue 5 and I had programmed to listen for cue 5 but he was actually sending cue 5.0. Once

I figured that out and updated my numbers, everything fired as expected.

On the flip side, there are times when you want to send a specific midi message to test why a device is not responding. You can use Midi-Ox to create a midi message and send it, but that is not always practical. Another option is the fantastic dBbox2, which can send certain types of messages. There is also Masque Midi Tester (Figure 18.10; www.masquesound. com/product/masque-midi-tester/).

Figure 18.10 Masque Midi Tester.

The Masque Midi Tester is a very simple device that can send Program Changes or Note On/Off to one of 16 midi channels. It is a nice single purpose device. I've had one for at least 15 years.

Video

I am a sound guy. Why am I dealing with video? Why? I guess because it falls under communications, and we seem to be responsible for that too. Video is a necessary evil in our business. Typically, every musical I have ever done has at least four basic camera shots:

1   Front of House—Color
2   Front of House—Black and White and Low-light
3   Conductor shot
4   Specialty shot

The two front of house shots are typically in the same location, which is usually on the balcony rail. These shots are for the stage manager and backstage crew. Sometimes the color shot will feed lobby monitors as well. The black and white shot is for when it is dark onstage or during a blackout. We hang an infrared emitter over the stage and that allows the stage manager and backstage crew to see in the dark. That way they can tell when the scene is set and the lights can come up. The conductor shot typically feeds video monitors on the balcony rail as well as monitors backstage and video monitors in the pit so that the cast and musicians can see the conductor and stay on beat. The specialty shot is usually for the stage manager or automation

Figure 18.11 Example of a CCTV tester.

or fly person so that they can see something important to them.

It is a good idea to have a CCTV tester in order to test your video signal in different areas. Just like midi, without this tester all you can do is test the cable, which may not be helpful. There are thousands of CCTV testers on the market. The most important aspect to me is that it be battery powered so I can walk around with it. Figure 18.11 shows an example.

A CCTV tester has a small video screen built in that allows you to see the signal. Some of these testers also have PTZ (Pan-Tilt-Zoom) control ability so that you can test a camera that has this function.

RF

When it comes to testing RF, this is a dark deep hole that we could get lost in for days. I have already written a section on RF Intermodulation, so I won't focus on that. Instead, I am going to focus on the mic elements. It is very important to test the mics. There are discrepancies between new mics. I have found the Sennheiser MKE-1 mics can be plus or minus 6dB from each other fresh out of the box. When I use MKE-1 mics on a show, I A/B test each mic and mark them on a scale showing what the level of the mic is. I mark the mics as "Loud," "Normal," or "Quiet." When I lose a mic on an actor, I can then replace with a similar mic. If I replace a "Quiet" mic with a "Loud" mic, then my input signal could be 6dB louder and there is a chance I could bring up the fader to my normal level and go straight into feedback.

In order to efficiently test and A/B lavaliere mics, there is only one product that I know of and that is the Masque Mic test box (Figure 18.12; www.masquesound.com/product/masque-mtb-51e-mic-test-box/).

Figure 18.12 Masque Mic test box.

This box has connectors for the most common wireless transmitter connectors, which are the Lemo connector for Sennheiser and the TA4F for Shure. This is a simple box where you plug one mic into an A connection and one mic into a B connection. You use headphones to listen and push the Selector button to switch between mic A and mic B. By using this tester, you can tell if your mics sound the same. Over time mics degrade. Some mics will lose gain over time. Other mics will lose high-end or low-end. You can also test mics for crackles and breaks in the cable without having to turn on the wireless. As with every other troubleshooting issue, this box allows you to eliminate all variables except the lavaliere.

Another issue that needs to be understood and tested with RF is your antenna diversity. Professional quality wireless mics have antenna diversity, which means you can hang two antennas and the receiver will switch to the stronger signal between Antenna A and Antenna B. One way to test antenna diversity is to have someone put on a wireless and walk all over the theater. On the receiver, you should see the Antenna meter bounce back and forth between A and B, showing which antenna is being used. If the receiver doesn't switch, then you need to ask the person wearing the wireless to walk to the other side of the theater from where the antennas are. It is possible that the signal is just solid and strong to one antenna, but that is rarely the case. We want them to walk away from the receivers because the receivers can pick up signal without an antenna plugged in. Distance will lower the signal strength. In fact, signal strength follows that good old inverse square law, which is double the distance is half the power. Once the person has walked away, then you unplug the antenna from the receiver that is being used. You should see the other antenna take over and have just as strong a signal. If you do not, then you have a problem with your antenna, the cable, or the antenna input on the receiver. To check all of that, you swap antenna A and B until you have determined the culprit.

Antenna placement is also very important. You never want to put an IEM antenna too close to an RF antenna. Why? Because an IEM antenna is broadcasting. An RF antenna is waiting to receive. If the IEM is too close then it could overpower the RF antenna with its broadcast signal. It is also good to get some distance between your RF antennas. If you have to put your two antennas side by side, then one antenna should be at a 90-degree angle from the other. The idea is that the difference in angle will allow the antennas to pick up transmissions that may have bounced off an object. I prefer to put my antennas on opposite sides of the stage.

Remember that the loss of RF signal through air is drastically worse than over cable. If you use RG-8X cable and run it 50' to the other side of the stage, then the loss of strength of an actor standing 1' from that antenna is about 5dB, but if both antennas are on the same side and the signal has to travel 50' to the antenna, then the loss of strength over that distance is about 40dB. That is significant. The lesson here is that it is better to run longer cable and get your antennas closer to the transmitters. Some people prefer one upstage and one downstage. I prefer one left and one right. My assumption is that the furthest an actor can be from either antenna would be up center, and they would be equally far from each antenna. The more they move in any direction, they get close to an antenna and their signal strength goes up.

It is possible to add an antenna amplifier to your RF system. If you do, then keep in mind that what you are trying to do is amplify the signal to make up for the loss of signal over the length of cable. So if we ran a 50' cable and lost 5dB of signal strength, then we would want to add 5dB of gain to the antenna amplifier.

Intercom

I have saved the best, or worst, for last. Intercom, or com. Let me string together some expletives to express how I feel about com. Actually, maybe I shouldn't or this book might be marked with a "Warning: Adults Only" label. Almost all intercom in theater is made by Clear-Com, and everything I will discuss here is based on working with Clear-Com. There are wireless systems as well, which I will discuss at the end. I have spent so much time in my career dealing with intercom. I have hunted down buzzes the way Elmer Fudd hunted wabbits. And I hate every single second of it. I know I can mix or design a great sounding show only to have the entire creative team and crew mad at me because of a buzz in the headsets that started in the middle of Act 1 and continued until the last scene in the show and is now gone. As much as I hate it, it is absolutely crucial to understand and know how to troubleshoot it.

The first thing I am going to make clear about intercom is that . . . are you ready? . . . Write this down . . . COM IS NOT SUPPOSED TO BUZZ. There, I said it. I have walked into so many theaters and asked how com was and have been told something like, "It's pretty good. There is just a little buzz in it." Well, guess what? COM IS NOT SUPPOSED TO BUZZ. But getting a com system to not buzz is a tedious and slow process. So where do we start?

A normal intercom system for a Broadway show consists of eight channels of com, which are: Deck/Carps, Lights, Spots, Sound, Lights Private, Moving Lights Private, Spots Private, and Sound Private. The private line for Sound is so that the mixer can talk to the deck sound people without anyone else hearing them. The three lighting private lines are so that the Lighting Designer can talk privately to the Moving Light Programmer while the Assistant Lighting Designer talks privately to Spots. The Lighting Private is a community private channel for Lighting. The fourth channel on any lighting com box would be the Lighting Public channel so that lighting can talk to the stage manager. This, of course, is constantly growing like everything else in theater. Now we have Projection and Projection Private channels. There can also be Carp private channels. It's a lot. Figure 18.13 shows the layout of a com system I did for the *Sound of Music* tour.

I use the word "Universe" to group together people who get the same feeds, such as the "Lighting Universe." All of these people are on the same channels. I decided to go with the word Universe because I felt it was unfair that the lighting world had confiscated the word. I still don't know what the lighting universes do, but it sounds cool when they say they have six universes in their system. I wanted to sound cool, too, so I started saying my com system had four universes. I have no solid data on this, but I am pretty sure it does make me sound cooler.

Once you have planned a com system, built it, and plugged it all in, then it is time to test is. The first step, and I can't stress this enough, is not to put on a headset and push the "Talk" or "Listen" button. The reason is because if that com channel is not terminated then it will squeal extremely

Figure 18.13 Example of a com
system layout.

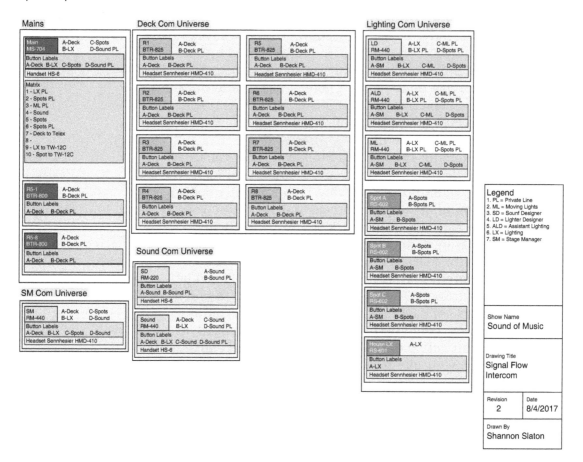

loudly in your ear. By not terminated, I mean that the com
channel is not connected to anything else in the system. If
you have a four-channel com box and you accidentally only
plugged in three XLRs, then one channel is unterminated
and will squeal if you turn it on. There is a trick to knowing whether a channel is unterminated, and this trick is the
first thing you should do whenever you walk up to test any
intercom device. That trick is to push the "Call" light for
each channel. If the channel is terminated and normal, then
the light will blink once or twice and then stop. If the light
blinks repeatedly for more than two blinks, then you can be

sure that the channel is not terminated and will squeal if you turn it on. If the call light turns on for one long solid blink that slowly fades out, then it means you have cross-patched a channel somewhere in your system, and if you talk into that channel it will be really loud and everything will be distorted. If you turned the "Talk" and "Listen" on it might squeal, but it is a different squeal that comes from the mic feeding back into the ear piece. Still unpleasant, but not as bad as an unterminated squeal.

Once you have tested the termination for each channel on the unit you are testing, you are ready for step two. Step two is to turn the volume all the way down and turn on the "Talk" and "Listen." You then slowly turn the volume up while talking. There should be no buzz or distortion. Just a clean voice. Why am I being so detailed about this? Because we are sound people, right? We make our livings on our ears, don't we? You could seriously do damage to your hearing by blowing your ear out with an unterminated squeal. Burn these first two steps into your mind. Make it your routine. It has to become second-nature because when you are in the weeds and trying to troubleshoot quickly it is very easy to accidentally skip these steps and blow out your ears. If this is 100% second-nature to you, then you have a better chance of doing it correctly while being yelled at by a carpenter in the middle of the show as you try and fix a com problem.

Now that you have checked the units to know that they are clean, you need to confirm the right channels are going to the right places. You are not going to make friends if stage management is getting the lighting private channel and the stage manager gets to hear the private jokes of the lighting team has about stage management. By the way, talk on com like everyone can hear you. In order to test the com channels, you need one person at the main station and another walking around to the different positions. When I toured I preferred to put a local sound person on the main station and I would walk to all the com positions. I know exactly how the system is wired so I can control the test. I tell the person on the main station that I will flash the call light and they will talk to me on the flashing channel. If I am at the Lighting Designer's com, I flash the Lighting, or LX, channel. The person on the main says hello and I ask them what

Figure 18.14 Masque
Clear-Com terminator.

the label says on the channel that flashed. If they confirm that they are talking on the Lighting channel, then I know we are good. I then tell them to talk to me on the next flashing channel and call them on the next channel on the unit. I make them confirm what the button says for each channel. I then move to the next unit and repeat.

At this point, you have confirmed your termination and wiring. It is possible that you have a com channel that is intentionally unterminated. The termination for a com system is the Main Station. If a channel is not plugged into the main station, it is not terminated. But there are times when you need to leave a channel unterminated from the main. By doing this you can create a "private channel" between units. If no one else needs access to the "Projection Private" channel, you could create a "Projection Private" between just two boxes in the house by connecting the two boxes on that channel without connecting that channel to the main. Now, this channel will squeal because it is unterminated. In order to make this work and create this closed-loop private channel, you will need to plug a Clear-Com terminator into the output of that channel on one of the two com units. Masque makes a Clear-Com terminator (Figure 18.14; www.masquesound.com/wp-content/uploads/2015/09/com-t-imagev2.jpg).

Another issue you can run into with intercom is a lack of power. Com works by transmitting power from the main station to the units. A main station is only made to power a certain number of units. You can find this information in the manual for the main station you are using. If you put too many remote stations on the main and draw more power than available, then you will trip the main station and put it into fault. You either have to hit the button to reset it or power cycle the main station when this happens. You can tell when a main station is overloaded because you will hear a tiny whine that fades in and out when people hit the call light. Your system could seem just fine even though it is overloaded. You could use it for days with no problem. That is because the overload occurs when multiple people key-in and hit the call light. Once that happens, the system will shut off. If this is happening, your only options are to get a larger main station, meaning go from a four-channel to a matrix main station. Or add a second main station to your system, which will add more power to your system.

So you have done it. Congrats. You have tested your system and it is rock-solid and silent. You are ready for tech. Then you come in the next morning and turn on the system and the world comes crashing down and there is a buzz. Buzzes can come from different places. One place is from cell phones. If I cancel out a missed pick-up for every time someone's phone caused noise in the com system, then I would officially have never missed a pick-up in my career. A bad cable could cause a buzz. Another place buzzes come from is the Clear-Com unit itself. And then there is the dreaded rack rails buzz, which I just had to deal with yesterday. At this point in the troubleshooting process for com, it is possible to have a completely silent system except for one buzzing box. To test this box, you should grab a spare box and test the cable to the test box outside of the rack. If you plug a different box in outside the rack and it is noisy, then your problem is a cable problem. If it is silent, then you are either dealing with a com unit that is buzzy or rack rail buzz. If the unit outside is silent, then replace the unit in the rack with the tester unit. If the tester unit is now in the rack and silent, then your problem is a bad com unit. Send it out for repairs. If the tester is now buzzing in the rack, then you are dealing with the Beetlejuice of com buzzes. An annoying little bugger that will be nearly impossible to fix.

The com system I am using this week is completely silent except for the stage manager's unit, which is buzzing. We narrowed the problem down to the rack rails. If we disconnect our power and video and announce connections to the rack then it is silent. If we plug any one of those three back in then it is noisy. The solution right now is we added some foam on top of the com unit so it is not touching anything metal and we removed the rack screws and wrapped the metal ears in tape and secured the unit with plastic zip ties. The unit is completely silent. We are still trying to figure out the culprit. Something has dropped ground and we will find it. No matter how long you do this or how good you get at troubleshooting com, there are still times where there is no clear explanation about why it is happening. Sometimes

Figure 18.15 ICOM-R Recording Adapter.

you can't fix the problem. When that happens, you have to find a way to avoid the problem, and part of doing that is remembering that COM IS NOT SUPPOSED TO BUZZ.

The day may come when your stage manager wants you to record her calling the show so that her assistant can learn to call the show. When this happens, you will need a DC Blocker. Masque makes a great one called the ICOM-R Recording Adapter (Figure 18.15; www.masque-sound.com/product/i-com-r/).

Using this adapter, you can plug a com channel into your favorite recording device. I usually put the SM calling on the left channel and put a program feed in the right channel. If you want to get really fancy, you can record the video at the same time by plugging the front of house color camera into your favorite video capture device.

Okay, so the next problem is one that pertains to the touring world. It's good to know but nothing you will ever need to know if you just have one com system in your theater. The problem is that touring shows travel with an entire com system. A touring show has to tie their com system into the house com system in most houses in order to get a headset to the person who will run the house lights or to the spot booth. Sometimes there is no cable path to get to those locations or it is just more work than necessary. The problem with tying in is that, in a lot of cases, the two com main stations are on different grounds and mixing grounds will cause a buzz in both systems. All of your hard work will get flushed down the toilet as soon as you plug into a house system. Sometimes the house will tell you they have dry lines to the spot booth and you will tie into those and avoid their com system all together, and you will still get a buzz. That is because those dry lines are sharing a house ground somewhere, which is different from the ground your system is on and that will cause a buzz. In order to eliminate this problem, you can use either a Clear-Com MT-701 (Figure 18.16; www.clearcom.com/product/party line/accessories/mt-701) or a TW-12C (Figure 18.17; www. clearcom.com/product/partyline/interfaces/tw-12c).

These units are isolation units for Clear-Com. The MT-701 is a single channel unit that completely blocks the transfer of power from the touring system to the house system.

Figure 18.16 Clear-Com MT-701.

Figure 18.17 TW-12C.

Both the house and the tour have to turn their main stations on. The result is no buzzy. The TW-12C is the same thing, except it can also convert different types of com systems to be compatible and you can control the volume to and from the house com system. I learned about these 20 years ago when I toured. After spending countless hours trying to get house systems to work with me, I finally bought an MT-1, which was the predecessor of the MT-701. The difference was that it was just a PCB and I had to wire it up and put it in a project box that I bought from this amazing store that no longer exists called "Radio Shack." When I toured I, loved my MT-1 more than anything else on my system. It was my most valued possession. And I never send a tour out without two MT-701s or TW-12Cs.

And that's it. I'm done with com. I don't ever want to talk about it again. I don't ever want to troubleshoot it again. I don't ever want to see it again. I look forward to retiring from this business and acting out the scene from *Office Space* where they destroy the computers with sledgehammers, except I will have a stack of old com boxes. I need to stop writing now because we have a workcall and we are going to find the rack rail buzz in the stage manager com so I should get over to the theater.

Riddle Me This, Batman

For the last section in this chapter, I challenge you to solve a problem. This is one of the most confusing and frustrating problems I have ever come across. I pride myself on my troubleshooting skills. I take troubleshooting very seriously, and I enjoy the slow methodological nature of it. This problem really pushed me to the brink, and I think it is a good example of why it is so important as a mixer to understand your system and troubleshooting techniques. I also think everyone should know that this problem can happen. I didn't and it really blew my mind. I will lay out the problem and give you opportunities to solve it, and then I will reveal to solution. Ready? Let's do it.

This summer I designed *Grease* at Maine State Music Theater up in Brunswick, Maine. It's a lovely little theater. I think it seats around 500 people. This is one of those summer stock theaters that slams a show in really fast but puts on really high-quality performances. They are like a well-oiled machine, as they say. The schedule is pretty insane. The last show loads out on Sunday night. Then *Grease* loads in on Sunday, while the other show is finishing loading out, and Monday. Monday night the cast goes into costumes and mics from 7–11pm and we tech through Act 1 with a piano. Tuesday we tech through the rest of the show between 1 and 5pm with the cast in mics and costumes. From 5–6:15pm we seat and sound check the band. At 6:30, without a sitzprobe or wandelprobe or even a soundcheck with the cast, we finish teching the show and do a full run from 8:30pm—11pm. Wednesday we rehearse from 10am—noon with the cast and have our first audience for a 2pm Wednesday matinee. It is crazy fast and you have to be on top of your game because the train doesn't stop. The producer is great. He understands how fast it is, but he still expects a great sounding show with no missed pickups for that matinee.

Grease went really well. We added the band Tuesday night and it actually sounded really good. I was pleased. The director was pleased. The producer was pleased. All around it was going extremely well. The band was nice and full. I had tons of gain before feedback. And even for the Wednesday morning rehearsal with just a piano, it sounded really good. I was relaxed and so was the mixer. We were smooth sailing through this crazy schedule. This was our third show for the summer so we had a system down and all things were happening as they should. We got to places and did the preshow talk. Handheld sounded great. Projection audio sounded great. Then it was time for the overture. We had been running the band fader between –5 and 0 for loud numbers so the mixer pushed the band fader up to 0 for the overture and instantly we heard feedback. Here is your first chance to solve the riddle.

Okay, I am assuming you haven't solved it yet. At this point, how could you? Nate and I looked at each other, just utterly confused. He brought the band fader down to –10 and the feedback went away, but the band was really quiet. And to me this is what being a true mixer is all about. The show started and we had a major problem. Our plane took off and basically the engine fell out, and we still had over two hours to fly that broken plane. As a mixer, you have to focus more on just mixing the show the best you can, and you have to find some way to mitigate the problem at the same time. People keep saying lines. You have to keep mixing them. You can't take a pause and work solely on the problem. You have to do both.

I was right beside Nate trying to figure it out. My first thought was maybe one of the Reed High mics was feeding back in to the front fills somehow, so I muted those mics. The next song in the show was not as big as the overture. It was "Summer Nights"; the feedback didn't happen for the first half of the song when we put the band fader at –5, as we had the night before. But all of a sudden it came roaring back halfway through the song, even though we hadn't changed anything. Nate pulled the band fader down and the feedback went away. It made no sense. Why would it come back out of nowhere? I then muted the Reed Low mics. The band consisted of two reed players, a brass, two keyboards, two guitar players, and a drummer. The problem didn't happen for a couple of songs and then it was back. I muted the brass mic and most of the drum mics. I knew it couldn't be the keyboards because they can't feedback and the guitars didn't have amps in the pit so that couldn't be it either. Yet it was still there. At this point, most of the band was muted and yet some songs would feedback and others wouldn't. Here is your next chance to solve the riddle.

After the show, we were exhausted from trying to make the show sound as good as we could with some random feedback. I was completely stumped. I told the director and producer what was going on and said we needed a one-hour soundcheck with the band before the evening show. Luckily, they had heard how good the show sounded the night before so they understood this was a strange anomaly, but they still were not thrilled.

The band took a one-hour break and I sat in the theater and looked over the console. I loaded the show from the night before and everything looked as expected. I brought the band fader up to 0 and there was no feedback. There was no one in the pit, but the system was solid. I could push the band fader to +10 without feedback. So why were we having problems? The band trickled back in to do the sound-check. I left the fader at 0 hoping I would find the problem when someone started playing. The drummer wailed away and there were no problems. Keys 1 and 2 played. The reed players and brass played. All was good. Then the guitar players sat down and instant feedback. Have you figured it out? I made everyone stop. Obviously, the problem was with the guitars, but why? The guitars were plugged into DI boxes and there were no amps. They were listening to themselves using Avioms with headphones. What was causing them to feedback when they were perfectly fine the night before?

I soloed the guitar channel and I could hear the feedback. I muted the foldback and the mains, thinking maybe somehow the guitars were feeding back in the speakers. Now there was no amplified sound anywhere in the house. I was soloing the guitar 1 input fader and I still heard feedback. It made no sense. Have you figured it out yet? The band fader was still at 0 even though the system was muted. I muted all of the inputs and still there was feedback. I finally brought down the Band DCA, which had been at 0. As soon as I got to around –10, the feedback went away, just like in the show. So, think about this. I am soloing an input fader. I am listening to it pre-fade. The Band DCA should have absolutely no effect on the signal I am hearing in my headphones. And, oddly enough, the Band DCA does not affect the level of the guitar I am hearing, but if I bring the DCA up to 0 then I hear feedback in my headphones.

My next thought was that somehow I had to be doing something that was adding signal to this input. I looked to make sure there was nothing inserted. I turned off the EQ. I then decided to look at what the guitar was being sent to in case it was somehow causing a loop. I was able to recreate this issue at will by bringing up the Band DCA to 0 so I asked the guitar player to not touch his guitar. I brought the DCA up until I heard feedback in my headphones. It was just a solid tone of feedback. I took the guitar out of the band group and then I started removing the guitar from all aux sends. When I removed the guitar from the Program feed the feedback stopped. That was the problem. The program feed was causing the feedback. I could turn the program feedback on and the problem was back. If the program feed was off, I could unmute everything else and the band could rock out at +5 on the Band DCA with no feedback. So riddle me this, Batman... how was the program feed causing the guitar input fader to feedback?

The answer to this riddle lies somewhere in the program feed. The program feed is a compressed full mix of the show that goes backstage and to the lobby and to the assisted listening. As Nate and I were talking this through, he realized that we had not turned the assisted listening system on until right before the matinee. That was the one thing that had changed between Tuesday night and the Wednesday matinee. But why would that matter? Well, this theater uses the new induction loop assisted listening system. Basically, the induction loop system broadcasts the program feed throughout the house, and anyone with a hearing aid can pick up the signal. This was our third show of the season. Why hadn't it happened before? It turned out that the company that installed the induction loop had, the week earlier, installed a stronger amplifier for the system. Electric guitars work by using magnetic coils to pick up the sound of the vibrations, and the inductive loop signal was strong enough that the guitar was picking up the program feed. Since the guitar was in the program feed, it created a loop inside the coils of the guitar. We tried unplugging our program feed to the assisted listening and plugging the theater's shotgun mic in instead, but that didn't help. If you soloed the guitar you could hear the shotgun mic as plain as day in the guitar. We had to turn the assisted listening off to solve the problem. The company had to come back and reconfigure the system. I can tell you it still isn't useable if you have a guitar in the show.

That night the show was back to normal and sounded great and everyone was happy. The lesson to be learned is that troubleshooting skills are crucial. Hone them. Be practical and methodical. Go slow and be deliberate. Eliminate as many variables as possible and build back up until you find the problem. And don't miss any pickups while you do it.

19

MIXING THE MUSICAL

"You don't have the emotional capacity to mix my musical."
—**Arthur Laurents, director and writer of *West Side Story* and *Gypsy*, giving a note to a mixer**

Well, here is the big chapter on mixing. I saved it for last because it is the most important, but also you won't get to this stage unless you do everything else. You have to plan the system and build it and load it in and program the console before you can mix. Once you have reached the point of mixing, then everything that preceded is in the past. You could've had the most perfect build and take-in, but if you blow the mix then none of it counts. If you had a miserable load-in and made huge mistakes but you mix an incredible show, then most of your mistakes will start to fade into the past. I spend a lot of time on every step that comes before mixing so that each part of the process is as complete as it can possibly be, so I have nothing to think about when it is time to mix other than mixing, because it is all about the mix. This is what we do. We mix. This is why most of the time I call myself a "mixer."

The problem with writing about mixing is there is no way to explain in words exactly how to mix. Mixing is like playing a musical instrument; there is no way you can write a perfect explanation of how to play the cello, any more than you can perfectly explain how to mix. What I can do, though, is explain some of the goals of mixing a musical and some of the tricks that are used to mix to achieve those goals. I can explain some of the sleight of hand and smoke and mirrors involved, and I can dispel some false assumptions about mixing and describe some habits that mixers can develop that will drive designers nuts.

First of all, being a mixer generally means feeling like a bit of a failure. If you are good at what you do, you will constantly critique your own mix and you will see all of the mistakes and flaws, even when the "mistakes" you hear are not even noticeable to an audience. Your ears will definitely be more critical than the average theatergoer's, but that doesn't mean we should be complacent about our mix. We can't lower the bar because we know our mistakes are imperceptible. Instead we have to always raise the bar, because even if our mistakes are not noticeable on a conscious level, they can be perceived on a subconscious level and distort the audience's response to the show.

Mixing a perfect show is basically impossible. I have mixed an average of 300 musical theater performances a year for the past 15 years and I don't think I have ever finished a show and thought, "That was perfect. There is nothing I could've done better." No matter how perfect your show, there is going to be a moment where an actor was too close to another actor and caused a change in the sound or a musician played something slightly differently and you didn't catch it or you were slightly too hot on a pickup. There are just too many variables to mix a flawless show. That is not to say that you can't walk away feeling good about your mix most of the time, but you will be a better mixer if you are highly critical of your mix. You will be in demand if you do not settle. You will excel if you demand excellence. I am pretty sure my grandmother had a pillow with that last sentence stitched on it, but no matter how cheesy it sounds, it is true.

When we tech a musical, we tend to get only a couple of passes at a scene and maybe a run or two before we have an audience. It is just the way it works. We sit in the dark for days as the lighting designer works his magic, and when he is ready we run a scene. If he needs to run it again, then we do. Otherwise we move on to the next scene. Most Broadway mixers take pride in being invisible. We don't want the audience to notice that we are amplifying the show. We dread being mentioned in a review, and we never want to hear the words, "Holding for sound," during tech. Our job is to nail everything perfectly every time we do it. In sound, we are only as good as the last show we mixed . . . the last scene . . .

the last song . . . the last word. Any flaw, and we are exposed like vampires caught in the sunrise. Lighting could lose a dozen lights at once and the audience might not notice but bring up one mic by accident and amplify an actor offstage flushing a toilet and everyone will notice.

With the schedule being what it is, this usually leads to a first preview that leaves the mixer feeling like he ran a marathon: drenched with sweat and exhausted from trying to mix a two- or three-hour show at full-speed for the first or second time and possibly with the cast finally singing full out, which blows any mix you thought you had nailed down. The result of this is usually a period of time where you just want to crawl into a hole and hide. The most important thing to remember is to make your pickups. If you make your pickups, you will mostly be forgiven for balance or EQ issues. If you make your pickups, the director and producer will give you time to get the rest of it straightened out, but if you are constantly missing pickups you are headed for, as my mixer friend Chad Parsley calls it, a "Come to Jesus" meeting with the designer. If you are missing pickups, people will go to the designer and complain and then the designer will come to you. There is a window when you will be ignored, which is before the audience arrives. Until there is an audience watching the show, sound will mostly be ignored, but once butts are in the seats, sound becomes of utmost importance. You have to realize that, as a mixer, your main job is to turn the mics on and off at the right time. That's it. If it doesn't sound good they will blame the designer, but if the mics aren't on they will blame the mixer.

So what do I mean by a pickup? The way we mix musicals is typically called "line by line" mixing or "one fader at a time." Our goal is to have as few mics open as possible at any given moment. If only one person is talking, then we only want one mic open. If only the guitar is playing, then we don't need the other 46 mics in the pit open. If three people are singing at the same time, we need all three mics open. We follow the script and we slap faders in and out as fast as we can. He talks. She talks. He talks. They talk. He talks. 1, 2, 1, 1&2, 1.

Why do we mix like this? Directors and producers ask this question a lot. I can't count the number of times I have been asked why we don't just turn all the mics on for the people onstage. Well, there are several reasons. The first reason is that it just doesn't sound good. If you have ten open mics and only one person talking, the sound of those open mics will make it harder to understand what is being said. It will change the sound of the room and make the speaker more distant. It will sound very similar to area mic'ing because an actor's mic picks up his or her voice, but so does the mic that is 10 feet away, which adds to the reverberant or ambient sound discussed.

The next reason is "phasing." Phasing is the sound of one voice being picked up by two microphones and being amplified. Technically what is happening is that the voice is being picked up by each mic at different times, depending on how far apart the actors are from each other. The waveform from each mic is being pushed through the speaker and because they are slightly delayed from each other, the result is the signal has some comb filtering as it is being amplified, which causes certain frequencies to be altered, thus changing the sound of the actor and bringing attention to the sound system as the actor's voice alters. When two sound waves are exactly the same, the result is a double in volume, but when two waves are exactly opposite, the result is the waves cancel each other out. If two waves are slightly different, the result is certain portions of the wave are cancelled out. As the actors move closer and further from each other, different frequencies get cut and then restored, creating thin and hollow moments.

There are several theories on how to fix this problem. The only one that truly works is an A/B sound system. An A/B sound system is a system where there are two speakers for every speaker location and the two speakers are focused on the same location. If you have an A/B system, which is very expensive since you need two of everything, then you put one actor in speaker A and the other in speaker B. This eliminates the comb filtering because the delayed waveforms are being produced by different speakers. Even when mixing on an A/B system we try to mix line-by-line, but if we need to leave two mics up at a time, then we can do it without altering the sound and drawing attention to ourselves. Mixing on an A/B system can be challenging because sometimes you leave mics open to create an ambience to the show instead of just for specific lines. When I mixed *Jersey Boys*, which I could easily argue was one of the top three sounding shows I have ever heard, I had notes on every line in the script about which mics were opened and which were closed because the sound designer, Steve Kennedy, was creating an overall sound of the space that he wanted to remain consistent. It was very effective and definitely not something that could've been done without an A/B system, but it was hard as a mixer because there sometimes was no obvious reason for why mics were open. It was a really enjoyable challenge to mix.

Another theory is to delay certain mics to eliminate the problem. This doesn't work because the delay only changes the frequencies that are affected. Another theory is to flip phase or, more appropriately, polarity on certain mic channels to solve the problem. The idea here is when the mic is polarity inverted, it pushes the speaker cone when it normally pulls it, and vice versa. This rarely works, and if it does, it only works for that particular moment, in that particular position, which changes as soon as the actor moves his or her head. I tried this on a really impossible moment in *Anything Goes*. I couldn't find any other way to make this one moment work, and it did work for about four shows and then the actors changed something and it stopped working and I had to find a new method, which in the end was a really complicated minute of mixing, but mixing is better than gimmicks any day. More often than not, it is going to make the problem worse as it completely hollows out the sound of the actor. If it does work, it is because the level into each mic is almost identical and the distance of the mics from each other is exactly right. If the level of signal into one mic is too different from the level into the other mic, the result will be to completely hollow out the sound of the actor. If you use this method, you are completely at the whim of the actors. If they vary their position or volume, you will be screwed.

Another reason for line-by-line mixing is for troubleshooting. The fewer mics you have open, the easier it is to find a problem. If you have eight mics open and you hear a crackle from a bad connector you are going to have a hard time figuring out which mic is bad, but if you only open one mic at a time you will instantly know who the problem is. Also, it is hard to mentally keep up with the show and what fader is important if you have several mics open all the time. The result is having an unbalanced mix where people are too loud or too quiet because you are never really sure who is speaking, so you don't know what fader to push.

Part of the challenge of mixing one mic at a time is figuring out how to deal with people singing or talking into each other's faces. A classic example is a duet between the romantic leads in a musical. Inevitably, they are going to start singing far away from each other and end up singing at the same time with their noses practically touching and ending in a big kiss. As they start to get close to each other, there will come a point where you will start to hear phasing. As soon as you get to that point, you have to figure out how to get rid of it. The best way to do this is to use one mic to pick both of them up. So as they get closer, you fade one mic out and avoid the phasing. A good rule of thumb is to use the male singer's mic or the taller person's mic. The reason this tends to work better is because, if you use the mic of the shorter person, the taller person is going to sing right into the other's forehead and be much louder. This doesn't always work, but it is a good first guess. Take time in rehearsal to experiment with which mic works best for these moments, and before you go grabbing for the EQ to fix the sound of an actor, make sure phasing isn't the problem, or the actor's voice reflecting off an object. Also make sure the mic placement is good. The best way to fix a sound problem is usually to adjust the mic position. Mixing to avoid this phasing as the actors move in and out of each other's faces is an exhausting game, but it is the key to a good mix.

Next it is time to learn the "Broadway Bump." When we mix the band, our goal is to follow the lead of the band and expand the band's dynamic range. When the band crescendos, we want to follow the band and crescendo with them. When the band pulls back, we want to pull back with them. When the band does a single quarter note bump, we want to bump with them. We want to follow their timing as they make these changes so it feels natural. If we fade up faster than the band, it will sound like two distinct fades. If we fade down slower than the band, it will be painfully obvious. Orchestras play with dynamics, but their dynamics have to be enhanced to make them accurate for the size of house, which is why we run the band that way. And, finally, we get to the "Broadway Bump." At the end of almost every song in musical theater is a button. Our job is to give that button a little bump at the end to accent it. The old "Broadway Bump."

There are several things mixers might do that will annoy designers. The first is to show off and miss a pickup. Nothing will drive a designer crazier than seeing a mixer mix a scene without a script and mess it up, or to see the mixer mix three characters line-by-line with three fingers on one hand while the other hand does nothing. The classic position for the Broadway mixer's hands is both hands on the board, with one index finger on an open mic and the other index finger on the mic that needs to come up next. I have been known to tell mixers, "I am not impressed with your one-handing mixing. I am impressed with a good mix." Another thing that can annoy a designer is excuses. It can get tiresome constantly hearing an excuse for why this person was too loud or why you missed the same pickup for the third show in a row. The audience doesn't care that the actor was singing quieter or changed his timing and they don't care that you mixed that scene perfectly yesterday. Our job is not to find excuses for our mix, but to mitigate the variables in our mix.

Another annoyance is over- or under-compression in a mix. When you are mixing vocals, your fingers are like little compressors. In fact, I will note my mixers by saying, "Can you please turn down the attack time and increase the ratio on your fingers?" That is my way of explaining what the goal is when someone is singing, because that is what we are doing. If a singer holds a handheld mic and sings, the singer will pull the mic away from her mouth as she sings the big loud note and pull it into her mouth as she sings the quiet note. In musical theater the mic is in a fixed position so the actor cannot do that. Instead, that is what we do. We compress them when they get loud and we open them up when they get quiet. If you under-compress them, the big notes will hit the audience too hard, and if you over-compress them they will feel squashed and buried in the music.

Mixers can form bad habits and hopefully a good designer will notice a mixer's habits and help the mixer find a solution. One bad habit is getting upset with your-self. Missed pickups happen. Your mind can drift and you can miss something. When it happens, you have to let it go and move on. If you don't, it can cause you to make more mistakes. I have watched a lot of mixers miss a pickup and quietly cuss themselves out and then miss three more pick-ups because they feel out of sync with the show. Kick your-self when the show is over. Figure out how to not mess it up again at another time. Break the habit of punching yourself in the dark while you try to mix.

Consistency is crucial to mixing. Our jobs are infinitely repetitive. It's the same script and same music night after night. The more consistently you make your mix, the better your mix will be. There is choreography to mixing a show and it is important to develop this choreography. Always turn the page of your script at the same time with the same hand. Create a rhythm for how the show mixes. Do it the exact same every time. By doing this you develop muscle memory and when you need to troubleshoot a problem your body can take over and go into autopilot and mix for a while, so you can try and figure out what the problem is and how to fix it. There are times when you have to mix a scene while talking on com to the A2 to explain a problem and how to fix it. If your mix isn't routine, you will have no chance of managing a crisis without the audience noticing.

Finally, we have to learn to mix the intention of the moment. This is where we become part mixer, part director, part musician, part artist, and part actor. We can't mix the entire show at one level. It has to have an arc. It has to have peaks and valleys. There are times when we want every-thing big and loud and there are times when we want to be extremely quiet. We have the power to pull people into the scene or push them away. Our ultimate job as a mixer is to make the audience laugh, cry, and clap. As a mixer we have to be painfully aware of the audience's reaction. We should be able to judge our mix by whether the audience laughed at a certain joke or applauded at the right time in a song. There is nothing better than figuring out a way to mix a song and milk applause out of an audience. I have told mixers that they haven't mixed a show correctly until they have made themselves cry—until they have become so in-tune with the show and the emotion of the show that they found a way to make themselves have a visceral emotional response to the show.

I can't explain how to mix because every moment is different and different shows require different techniques, but I can tell you that our goal is always the same, and I can tell you that you will never achieve the goal unless you mix line-by-line. You will never make an audience cry unless you invest yourself in your mix, and you will never be a good mixer if you accept anything less than perfection and you understand it is not possible.

20

MY ADVICE TO MIXERS

As my career has progressed, I have followed a well-trodden path of past Broadway mixers and have found myself in the position of designer more and more. In fact, I am actively trying to avoid mixing jobs to make myself more available for design jobs, but just as Al Pacino says in *The Godfather Part III*, "Just when I thought I was out, they pull me back in." If you haven't seen that movie, then go watch it now. I'll wait.

Currently I am mixing monitors for Bruce Springsteen on Broadway. That's right. I am mixing for the Boss. Brian Ronan called me to ask if I was interested. He knew I was moving away from mixing, but he thought I might be interested, and of course I was. I mean, I would prefer to design now, but I still love mixing, and come on. Who wouldn't want to do this gig for a couple of months? Of course, it has led to a crazy few months for me as I had already booked a lot of design work. In three months, I designed the tour of *Sound of Music* and *Amazing Grace*, two *Elf* tours and *I Love You, You're Perfect, Now Change* at a regional theater, a one-night high-profile show at Town Hall, and four readings—and I was the associate for a Broadway play called *Meteor Shower*, written by Steve Martin and starring Amy Schumer and Keegan-Michael Key. I somehow managed to do all of that while building, loading in, teching, and mixing *Springsteen on Broadway*. Sometimes a career in this business feels like juggling cats on no sleep.

But the point is that I am making the transition from mixer to designer and find myself working with a lot of young mixers. I find that there are lots of things that I say to every mixer I work with. It's basically the best bits that I collected in my bag of tricks that I find helpful for young mixers to learn.

I have found there are some basics that I know that can very quickly give a young mixer enough experience points to move up a level and add a proficiency bonus (that is a reference to "Dungeons and Dragons," which is by far the greatest game ever created. My son and my friend Mike have a podcast called *My Son the Dungeon Master* where we play D&D.) I thought I would put some of my best tips here. Some may be painfully obvious or already common knowledge to you. If so, congrats on being a fourth-level elven halfling rogue, which is the only race and class for sound people. No matter how commonplace these tidbits may be, hopefully someone out there will get something out of them.

The Love Duet

If you are mixing musical theater, then odds are good you are dealing with a love story. And if you are dealing with a love story, then you are going to be mixing a duet between the lovers at some point. And if you are going to be mixing a love duet, then you are going to have the lovers close to each other, or in each other's arms. They will be singing back to back. Front to front. Side by side. Front to back. They will run the gambit of the musical theater Kama Sutra blocking. In all of this movement, you are going to have to make dozens of decisions about how to mix this song without hearing phasing, which is when one voice gets into two mics and causes cancellation of some frequencies. So what do you do? Is there a guide or a trick to this?

Of course there is. Why else do you think I am writing about this? The simple rule of thumb to go with is to kill the mic of the shorter actor when two actors are singing into each other's faces. If it is a duet between a man and a woman, odds are good that the woman will be shorter. It is just part of the unconscious societal programming that we live with that leads to casting choices of women shorter than men. There are several reasons to kill the mic of the shorter person. The first reason is because the taller person will be singing directly into the forehead of the shorter person, and if you try to use that mic the you are going to get a lot of the taller person. The second reason is because the shorter person probably has a higher singing voice that lends itself better to being picked up farther away. A third reason is you are less likely to pick up string plosives. So, if you see the couple getting closer as the song nears the end, then the first thing to try is to dump the shorter mic.

Your job isn't done at this point. You don't get a trophy for making this amazing choice. Now you have to look up from that script and pay attention to what they are doing. This isn't just mixing. It's choreography. They pull in close to sing and then pull away. Back to close and the woman turns her head. Then they are both looking straight out into the audience in a tight embrace, and then they turn toward each other and go in for the kiss on the last note. You have to do the choreography with them. Now you need her mic, then both; now just his, then both, then his, then hers, then his and kiss. When done correctly, you can mix the song using different combinations of mics with no phasing, and the audience will have no clue you are in the back of the house throwing faders to gently balance this mess.

The Dead Mic

This is the nightmare we all live in fear of. We have a million-dollar sound system with the latest and greatest of everything, and it is all dependent on this tiny cable not failing. And we all know that lavalieres fail. Of course, we don't just have one of the tiny cables in a show. Where would the challenge be in that? No, we have thirty of these mics onstage at any given time just ready to fail. It's no wonder why sound people tend to be more superstitious than other disciplines. "I've never had a mic fail when I mix wearing these shoes."

Part of being a good mixer is having an emergency plan for everything. Assume the worst. The first thing you need to look at is who should be wearing two mics. Certain people are onstage so much that it would be almost impossible to fix a broken mic. Those people really need to be wearing two mics and you need to come up with a plan for how to swap between the two mics seamlessly and maintain your console programming for the remainder of the show. The easiest way is to patch the new mic in where the bad mic is. On a DiGiCo you can do this very easily because every input has an A and a B input patch. I wish all consoles had this feature, but I could say this about dozens and dozens of features of the DiGiCo consoles.

When a mic fails, because it will, it is important to mitigate the disaster as much as you can. There is no getting the egg off your face. You just have to make is sound as good as you can. First thing to do is bring the level of everything down. Make the moment as acoustic as you can. Second thing to do is watch the stage. If the actor with a bad mic is near another actor with a good mic, then try using the other actor's mic to pick the person up. There are lots of actors who will realize there is a problem and they will look for someone to stand next to. There are other actors who will realize another actor has a bad mic and will follow the actor around as basically a boom mic operator for you. The most important thing to do is to communicate with backstage to make sure there is a plan ready to go to get the problem fixed.

Keep the Designer Informed

It is always a good idea to make sure the designer hears about a problem from you before they read about it in the performance report. You never know how the stage manager is going to explain a problem. I remember a show where the music director triggered some click tracks and occasionally when a sub conductor was, in the triggers would be obviously late. Unfortunately, the stage manager couldn't seem to understand that this was a mistake by the music department and would instead write in the report that Sound fired the cues late. My mixer contacted me to explain what happened, so I would respond to the stage manager and everyone else that this was not a sound mistake. We have enough real mistakes and problems in sound from missed pickups and broken mics; the last thing we need is to be blamed for something that we didn't do.

The whole point is that if you keep the designer informed, then the designer can be an advocate for you as a mixer. If you do not do that, then you may find that a performance report went out and the producer called the designer to figure out how to get a new mixer. I've had this happen several times. If I know the backstory of what really happened, then I can protect my mixer. One perfect example is a tour I designed of *Sweeney Todd*. The tour had been out for months and had gotten nothing but rave reviews. Then one night my mixer, Jacob, called me at intermission in a panic. He told me the show sounded really bad and he couldn't figure out why. He said it sounded good at sound check but as soon as the show started he knew something was wrong.

Jacob was mixing from a bad position in the back and under a very deep balcony. He said it seemed loud and cavernous in front of him, but he couldn't figure out why. I told him to mix the second act quieter and to call me after the show. I immediately sent an email to the producers, general management, production manager, and stage manager. After the show, Jacob called me back and told me Act 2 was worse. He said the volume seemed much louder and it sounded like things were distorting. He said he tried turning the show down but it seemed to just get louder. I told him to try and figure out what happened and then call me back. I sent another email saying the show sounded really bad and that we were trying to figure out what happened.

Jacob finally called me back and explained what happened. It turned out that the house technical director and house sound person did not think the show was loud enough during the sound check. Jacob had given a feed to the house for their center cluster and delay speakers and under balcony speakers. The two house guys decided to "fix" what they didn't like so they were walking around the theater with a computer turning speakers up. At intermission they received complaints so instead of telling Jacob, they decided to double the output of their amps. I sent another email to everyone explaining what happened.

It is important to understand that this is a business and our goal is to make money. The actions these two people did caused the theater to have to refund hundreds of tickets because of bad sound. Because Jacob had emailed me at intermission, I was able to show a paper-trail that we were aware of and working on the problem. When we found out what had happened, I was able to make the argument that this was not the fault of the mixer or the touring company. This was the fault of the people the house had hired and therefore the theater was responsible for the cost of the refunded tickets. The house agreed and there was a meeting the next morning. Those two people were not allowed back in the theater and the theater paid for a second sound check. We had no other complaints for the rest of the stay at that theater. Of note, the review came out based on the bad sounding show and it said my sweet little sound design sounded like "Norwegian Death Metal."

Levels and the KISS Rule

It is important to understand your system. What is your underscore level for the band? What is the level for a song? What is the level for a dance break? At what point on the fader does the band drop out of the system? Where does dialogue live on the fader? How about songs? These are crucial things to know. If you can't answer these questions because things seem all over the place, then your system isn't set up properly. You should know that it is safe to throw any actor's fader to –5 and it will be approximately a good level for dialogue. If you have some actors that are a –20 throw and others who are –5, then you are going to have a very hard time mixing that show. You need to balance the levels on either the input gain or the input fader. Same goes for your band. If your band is all over the place, then you need to work on that input mix.

If I can answer those basic questions, then I should be able to walk up to a show, cold, and be able to be in the ballpark. Our job is hard enough. All those fancy knobs and buttons on the console are there to make our lives easier, not harder. We are trying to make order out of chaos. I am never impressed with a one-handed mixer, unless you are a one-handed person, in which case I am highly impressed. I am impressed by a good, solid mix. If that means you mix with your feet, then bravo. But if I see someone mixing a scene with one hand and missing a pickup, I will lose it. We didn't evolve from two-handed monkeys just so that you could ignore one hand.

I live by the KISS rule, which is "Keep It Simple, Stupid." It seems like the more experienced a mixer is, the easier everything is. I think it is because you learn over time that complicated and fancy is unhelpful. When I started out I would have four faders for the band. I would break it up into sections like: Drums, Bass, Keys/Guitars, and Brass/ Reeds. Why not? I have four fingers and I am a really good mixer, so why not break it up? But over time I realized that it took more effort to manage the band like that for every song and it made it harder to focus on the singing, so as time went on I simplified more and more. I can't remember the last show I did with anything more than a Band fader. I will break it out for a song if needed or I will build a fader bank of the band inputs I touch for solos, but my VCA section stays simple. The same goes for singing. There was a time when I would break out Altos and Sopranos. It looked great when I mixed. Fingers flying everywhere. And I could do it, but as soon as a mic would break or anything else would happen I would crash and burn on the complicated mix. So I can't stress enough. Keep It Simple, Stupid.

21

TIPS FROM THE PROS

Bob Biasetti

I learned two shows from Bob. First was *Dirty Rotten Scoundrels* and second was *Legally Blonde*. Bob is one of the most consistent mixers I ever worked with. He was also never missed a pickup. I mean never. I strive to be as good as Bob. His passion and dedication to mixing is infectious and he always sweats the little stuff because nothing is little.

1. What Are Some of the Shows You Have Mixed?

The King and I (Donna Murphy, Lou Diamond Phillips)—4 TONY awards; *Annie Get Your Gun* (Bernadette Peters)—2 TONY awards; *Matthew Bourne's Swan Lake*—3 TONY awards; *Oklahoma*—1 TONY award; *Dirty Rotten Scoundrels* (John Lithgow)—1 TONY award; *La Bohème on Broadway* (Baz Luhrmann)—3 TONY awards; *The Boy From Oz* (Hugh Jackman)—1 TONY award; *Billy Elliot*—12 TONY awards, including TONY for best sound design

2. How Did You Get Started as a Mixer?

I started out as an assistant staff engineer in a recording studio. I started off working at the Annenberg Center at the University of Pennsylvania. One of the shows that came thru was moving to Broadway. I asked the sound designer if he would try and help me get the job, and it worked. From there I got a job at Promix (now PRG) and worked there for about 6 or 7 years. While there I took all the outside jobs I was offered. Some thru the shop. Some not. Eventually getting my first Broadway show.

3. What Advice Would You Give to a Young Sound Person Trying to Make It?

There is no right or wrong path to follow. Ambition is a good thing but take some time to enjoy the show you're mixing. Don't do what I did and always think about moving on to the next, the next and the next. Take some time and enjoy it. There will always be a next show.

4. What Was Your Favorite Show to Mix and Why?

That's an easy one. *La Bohème*. Hands down the best experience of my career. Getting these young opera stars to trust you while wearing mics was a big deal. To start, I do not speak Italian and I do not know how to read music. When we did production at the Curran Theater in SF, one of the first things I did was to buy a new laptop with a DVD drive. I would set up the laptop on the Cadac side car and watch the Australian production of the show scene by scene and use that to program the show just enough to stay ahead of tech. The added fun was that there were 3 sets of Rodolfos, Musettas, and Mimis. During rehearsals and tech Baz would swap them around randomly to work out the best pairings.

5. What Advice Do You Have for Working With Designers?

Always seem like you've got everything under control. Especially when you may not. Every designer has odd thing they care about the most. Even if it makes no sense to the crew, get that done. You'll be happy you did. Still to this day THE most important thing . . . DON'T MISS PICKUPS. That's the one thing the SD can't fix. The designer needs to feel like he can do his job. I always tried to make sure I didn't miss a single pickup on that first day. It sets the tone and lets your boss know he's in good hands. It may sound old school but it's still the one thing that you can't fake. It will get you fired.

6. Do You Have a Good Mixing Horror Story?

A big show that will remain nameless and a sound designer that will remain nameless. Some quotes from that SD as said to me. I swear.

"I need you to make it sound more BLUE"

"Take me on a journey"

"Here is something I learned when I was a mixer. When you push the fader up it gets louder and when you pull the fader down it will get lower."

You can't make that shit up!

FRANCIS ELERS

Francis taught me the mix to one of my favorite shows, *Spring Awakening*. He also made me reevaluate mixing. His style of mixing is very in the moment and passionate. Francis has so many good tricks I think I could write a whole book just about what he taught me. He will intentionally choose not to make something in the mix easier because he understands how easy it is to get complacent after mixing a show 500 times. His theory is that sometimes it is good to have to reach for a knob and turn it even though you could automate it because it keeps you in the moment. He has a good point.

1. What Are Some of the Shows You Have Mixed?

Broadway musicals include *RENT*, *Assassins*, *Spring Awakening*, *Into the Woods*, *42nd Street*.

2. How Did You Get Started as a Mixer?

By mixing things other than musicals, like concert tours and industrial events. Eventually I decided I wanted to try musicals and spent some time freelancing exclusively for Autograph in London. Bobby Aitken gave me my first show, a jukebox musical called *In the Midnight Hour*. I arrived in NYC from Europe in about '95 and Brian Ronan, who I'd met through Tony Meola (who I knew from a transfer of *Anything Goes* to London) very kindly found a slot for me at *RENT*.

3. What Advice Would You Give to a Young Sound Person Trying to Make It?

I was going to write "Try not to be too dismissive of us oldies just because we're not very good at configuring networks and we'll try not to drone on about the fact that you've never edited audio tape with a razor blade," but it's a serious question and deserves a better answer (albeit in the same vein). Competence is great but a little humility along with it will take you miles. I have in mind one of the younger people "on the scene" at the moment. He's super capable, not cocky in the slightest, always prepared to listen and accept advice, and he's head mixer at one of the current mega-hits. He absolutely deserves his success.

4. What Was Your Favorite Show to Mix and Why?

The Look of Love, a Burt Bacharach/Hal David musical. Two hours of Bacharach/David songs, wonderful singers including the incomparable Capathia Jenkins, exciting orchestrations by Don Sebesky and a lovely Cadac/L-Acoustics sound system by Brian Ronan. Oh, and no book scenes!

5. What Are Your Favorite Tips and Tricks as a Mixer?

Be wary of mixing music of a style that you don't listen to yourself often, by choice, for pleasure. This is not to say that I listen to musicals at home or in the car; I don't. But I do listen to a huge amount of pop, funk, soul, and jazz. I have, on occasions, mixed symphony orchestras (outdoor concerts when I lived in Germany, for example), and the experience makes me uncomfortably aware that I simply don't listen to this type of music often enough to do the job well. I've been listening to pop music for 50 years. I'm clear in my mind about how it should sound.

6. What Advice Do You Have for Working With Designers?

Develop an ability to work out quickly whether or not a designer is collaborative by nature and learn to roll with it either way. Some are, many are not, and if you want to keep working for the "nots," you need to accept that fact and learn to be OK with it. (And yes, I have discovered this by bitter experience.)

7. What's the Best Note or Bit of Advice You Were Given?

"If you want to be given the opportunity to do something (like mixing), don't be seen doing anything else." (© Tony Meola). To explain this a little: if people see you in a safety harness clambering around on truss all the time, they'll pigeon-hole you as belonging in that role. It's unfortunate, but inevitable.

8. Do You Have a Good Mixing Horror Story?

I did a Palestinian arts festival in East Jerusalem once. Audiences there have a habit of discharging their weapons in the air at the end of every song. That was pretty horrifying for a boy from rural England (but perhaps this wasn't the sort of thing you meant?).

KAI HARADA

Kai was one of the first people I worked with when I moved to New York. I learned a lot from him, including the importance of every detail. Kai cares about his designs down to the font used on a P-Touch label. He is always one or two steps ahead, which is really the only way to survive in this business.

1. What Are Some of the Shows You Have Designed?

Broadway: *Amélie*; *Sunday in the Park With George*; *Allegiance*; *Gigi*; *Fun Home*; *On the Town*; *First Date*; *Follies* (Tony Award & Drama Desk Nominations); and *Million Dollar Quartet*. Selected Regional/International: *A Legendary*

Figure 21.1 Kai Harada.

Romance and Poster Boy (Williamstown); *Beaches* (Drury Lane); *Brooklynite* (Vineyard); *Little Dancer* and *First You Dream* (Kennedy Center); *Zorro* (Moscow; Atlanta); *Hinterm Horizont* (Berlin); *Sweeney Todd* and *Pirates of Penzance* (Portland Opera); and *She Loves Me* (Oregon Shakespeare Festival). Audio Consultant for the revival of *Hedwig and the Angry Inch*.

2. How Did You Get Started as a Designer?

I started my career in professional theatre primarily as an assistant, working for renowned sound designer Tony Meola. [Over] the course of nearly fifteen years I assisted him on such projects as *Kiss Me, Kate, A Christmas Carol, Man of La Mancha*, and *Wicked*. My responsibilities grew with each show as we developed a working relationship, and by the time we got to *Wicked* my responsibilities became production sound engineer for the additional US companies. When there was time I started designing smaller shows, including Off-Broadway musicals and touring kids shows, and that, coupled with my experience mounting *Wicked* over and over, helped hone my skills as a designer. I only mixed a little bit on Broadway, but having spent so much time in theatres writing down mix notes on behalf of Tony, I knew what notes he would give; getting behind the console myself also made me a better designer since I can give more constructive mixing notes to a mixer. I am a terrible A2, but I hire people who are fantastic A2s.

3. What Was Your Path to Designing on Broadway?

A little show that we did in Chicago called *Million Dollar Quartet* ended up coming to Broadway, and that was my first Broadway credit. It was sudden and surprising but also very gratifying, and I was very pleased with the way that show sounded. Doing that show [I] also established a great working relationship with a director who did *Follies*, which I ended up designing as my second Broadway show.

4. What Do You Look for in a Mixer?

Musicality/talent, personality, and skill. As a designer the mixer is there to replicate my sound design, day after day, and we will spend a LOT of time together setting up the show, so it is really important that we *get along*. I have a very specific personality type which meshes with some people, and not with others, and I need to work with someone I can communicate well with; someone who hears similarly to the way I do; and someone who has an emotional connection with music. There are some mixers who are very good at making quick pickups and mixing a very fast show, but making pickups is only the first step: I care more about what happens once they've made the pickup—how they can help sculpt what the actor is doing vocally, and create an emotional dynamic for the audience—someone who instinctually understands musical arcs to help create the performance and doesn't just put faders up at the same level every time. As for "skill," I enjoy working with people who can troubleshoot and engineer other sections of the sound system well; a mixer's responsibility extends far beyond the console and programming the show, so a savvy sense of signal flow and logical thinking is important to me. I will often take two out of those three criteria, but never just one out of three.

5. What Do You Look for in a Deck Sound Person?

Similarly, personality, skill, and talent: eventually the deck sound person will learn the mix, so all of the above qualities apply, but it's also important that the A2 is personable, because they are the first line of communication with the cast and/or musicians, and has an excellent bedside manner. Speed and creativity [are] also key, as sometimes we are tasked with putting microphones in strange places. I appreciate a proactive A2 who helps hide microphones well (which is something I try to do on most shows) and keeps actors comfortable backstage.

6. What Advice Would You Give to Someone Trying to Make It as a Mixer or Deck Sound?

Do everything. I did some terrible rock and roll club work when I was younger, trying to make knock-off SM58s through a shitty twelve-channel console into terrible speakers that were pointed badly, and sometimes being able to work with the worst of the worst is a very important skill to learn. Starting out, you're not always going to get the best and newest equipment: make it work with a bunch of SM57s, some carpet-covered speakers, and an analog console; this is the best way to learn. Do some load-ins Off-Broadway and see how different designers work, install things, label things, etc., etc. Work on shop builds and do the same thing. If you're decent at building and installing, and have a good personality, you will get hired again, and eventually you can make the transition to A2 and A1. I meet a lot of young people who come out of school programs who just want to mix; unfortunately, that's not how this business works—while I don't want to sound like a typical old person talking about "paying your dues," there is an element of that as well as understanding that the more you know about ALL facets of the business, the better a well-rounded sound person you will be.

Like we talked about, I'm a big proponent of realizing that there is no single way of doing something in sound—sure, there are laws of physics and gravity and acoustics that are pretty darn immutable, but with the tools we have at our disposal in terms of equipment, there are tons of ways to solve any particular issue. There are definitely "traditional" ways of doing things, and a VERY loose set of standards for the New York market, but it's more of a standard of *what* we do, rather than *how* we do it. Do we almost always specify our own cable for a show in a particular New York theatre? Yes. Do we almost always label it? Yes. Do we all label things the same way? Certainly NOT. Do we always have to incorporate intercom and video into our designs, since none of those systems are built in to New York area theatres? Yes. Do we all use the same systems and set them up the same way? Definitely not. So, be flexible. Ask questions. Stop thinking that you know more than a seasoned sound person just because you went to some school.

7. What Are Your Favorite Tips and Tricks for a Mixer?

Go to the run through in the rehearsal hall with your script, and really pay attention to blocking, staging, and vocal dynamics. What actors give in rehearsal may be all that you get on the stage, so it's important to note when people will be facing upstage, screaming into each other's mics, putting on hats, whispering, etc., etc., etc. And then write down when the music is, so you know the show better than the designer. This isn't really a tip or a trick, but it's an expectation. Which is really covered in the next section.

8. What Do You Expect From a Mixer?

I expect a mixer to program the show, VCA-wise, to the best of their ability and comfort. I don't particularly care if the orchestra is on the first VCA or the last so long as the sonic result is achievable. I have many guidelines about what console features to implement or program while programming a show, but the actual assignment of VCAs is something I leave up to the mixer. The mixer should then be proactive about consulting with stage management about line changes, or when actors start ad-libbing, and pay attention to everything that is happening in the tech process. One quiet note from a director to an actor can totally change a line reading, so it's important to keep an ear out to what everyone ELSE is doing when the mics aren't up.

9. What Do You Expect From a Deck Sound Person?

The A2 should take initiative to keep the microphone in the position that we decide (often with the actor in a mic fitting session that I like to oversee) and keep the actor comfortable and confident that the sound team has their back. I like to remind my team that all of them are representatives of me, the sound designer, and attitude is everything—this goes for conduct backstage with other departments (like hair or wardrobe) as well as with actors and musicians. We are there to support the show, and my team is there to aid in that process however possible.

10. Do You Have a Good Mixer Horror Story?

Sometimes when I design regionally, I am obligated to use house staff (and/or house system as well). This is always something that makes me apprehensive, but often there is no recourse. This often means that I have to adapt to how a house staff works and adapt to existing microcosms of sound people and different personalities in order to get what I need out of the team. Doing a little bit of touring work as the Advance Sound Engineer helped me learn how to handle a different group of local stagehands and figure out how to get them to do what I needed them to do in a short amount of time. It's much the same when walking into a regional theatre with an existing house staff. So I designed a show that did a four-stop regional "tour" in a way; the actors, costumes, and scenery remained the same, while the lighting designer and I were obligated to use whatever house sound system and rep light plot existed. It also meant that, in spite of my protestations, that I would be obligated to use the house mixer for two of the cities. The show was a fast-paced pop musical, already programmed onto on type of digital console, so I demanded that we at least keep the console and its programming. We forwarded training videos and recordings and a[n] already-filled-out script and the console file, and yet, when we walked in for our first day of tech, it was a disaster—missed pickups, no idea about band levels, late sound effects; I think the poor kid just didn't know what or how to study using those materials, and it took a nasty meeting with the production manager to ask if 20% missed pickups was acceptable for a preview for them, because it certainly wasn't acceptable for me or the director. Eventually we kept our original mixer, who was there in an assistant role for me, through opening to mix the show, obviously at the theatre's expense.

I think a lot of regional theatres underestimate the magnitude of difficulty when doing a musical—in ANY department. I wish I had [a] nickel for every time I heard someone in a regional theatre say, "musicals are hard!" THEY ARE.

11. Do You Have a Good Mixer Success Story?

I took a chance on a young mixer who had just graduated an undergrad sound design program because I thought she had a good attitude and a decent set of skills; I was designing a small Off-Off-Broadway musical for a college friend and so I let her mix the show, with minimal tech time and rehearsals, and she programmed the show and mixed it extremely well. She's now doing quite well for herself, having done some touring shows and other things in New York.

JULIE SLOAN

Julie taught me the most difficult mix I have ever learned. It was *Jersey Boys* at the August Wilson Theatre on a Cadac J-Type and an A/B system. The thing that made it so difficult is that we didn't mix line by line in the traditional sense. We would leave some mics open even though the person wasn't talking so that the ambience of the space would not aurally change. It was a great effect and that show sounded amazing. It was one of the most satisfying shows I ever had the pleasure to mix. The thing of note with Julie is her unbelievable calmness. I can imagine her mixing a show with her console on fire and her casually putting out the fire without missing a pick-up.

1. What Are Some of the Shows You Have Mixed?

First National Tours of *Annie Get Your Gun*, *Aida*, *Jesus Christ Superstar*, and *Hairspray*; *Dessa Rose* at Lincoln

Figure 21.2 Julie Sloan.
Photographer: Tomas P. Wensel

Center; on Broadway, *Jersey Boys*, *Guys and Dolls*, *Jesus Christ Superstar*, *On Your Feet*, and *Charlie and the Chocolate Factory*.

2. How Did You Get Started as a Mixer?

I was part of the stage crew at my high school and enjoyed it. I did a year in college (Butler University) as a music major, then dropped out and waited tables for three years. I went back to college, as a music major again, and found that my school (Indiana University) had an audio program that was part of their School of Music. I applied to that program and was accepted and pursued both the BS in Music and the AS in Audio Technology and then, on a gig, had an accident with a pocket knife. Always use the right tool for the job, and always cut away from the body. The right tool for cutting zip ties is not a pocket knife. I cut through three tendons in my left hand, and I was a classical guitar major. After surgery and PT, I did regain nearly full use of my hand but had to face the facts: I was a pretty crappy guitar player. But I was pretty decent at this sound thing. . .

To complete the AS degree, I had to do an internship and got one at the Dallas Theater Center. While I was there, Arizona Theater Company offered me an actual paid position to mix one of their musicals and I jumped at the chance. I continued working in Arizona after that, mixing musicals but also working with shops, doing corporate gigs, doing rock and roll, serving as systems tech, etc. I also applied to the IATSE Local in Phoenix and entered as an apprentice. Shortly after that I was offered my first tour. One gig always leads to the next.

3. What Advice Would You Give to a Young Sound Person Trying to Make It?

Hmm. I'm gonna get ranty. You're probably gonna have to edit this. [I didn't.]

Most important: Don't be a dick. Even if you're the smartest and most talented person on the gig (and you're probably not), people don't easily forgive you acting like a jerk. Do not try to impress the old dinosaurs like me with how well you know the newest, flashiest piece of gear. I'm far more impressed by qualities like a strong work ethic, kindness, and a sense of humor. This is advice applicable to pretty much any industry. But sound in general, and specifically sound for theater, is a small business. We all know each other eventually. And we all talk to each other.

Most important for sound: I get that we're deep in the digital age. Sound is still all about signal flow and line levels. UNDERSTAND GAIN STRUCTURE. Today's consoles give us all kinds of amazing tools, but the integrity of your system depends on you adhering to the laws of physics. We're frequently running into young people coming up in the business that are probably otherwise quite smart and talented, but they fail at troubleshooting because they don't understand signal flow, and their very first tool to dial in a microphone is some kind of dynamic filter. If your grasp on signal flow and gain structure is weak, get your hands on some analogue gear and set up a system from scratch. Do it a lot. Fix the cables when they break. With actual solder. Listen to the difference between your preamps barely cracked, wide open, or hanging out in the sweet spot.

Trust me, you'll be light years ahead of that smart guy who knows the latest piece of gear and wants to tell you all about it.

4. Which Show Was Your Favorite to Mix and Why?

It's hard to pick just one. I loved *West Side Story*, which I did two local productions of in Arizona, because you pretty much have to mix the montage off the score. I loved the Broadway version of *Superstar* because our band was just on fire. Likewise, I loved *On Your Feet* for the same reason—most of the band was actually Miami Sound Machine, and it's an amazing experience when you get to mix guys that have been playing together for twenty years or longer. And of course, I loved *Jersey Boys* because I was with it from the beginning out of town tryout in La Jolla.

5. What Are Your Favorite Tips and Tricks as a Mixer?

In the upper right-hand corner of my script, I put the cue number that we're in on that page. When you're in production, this is super helpful when they bounce around.

I move towards getting off book as soon as possible, because a whole world opens up when you can actually look at the stage while you're mixing. That said, sometimes I close my eyes to listen better.

I spend a not insignificant amount of time practicing redirecting my mind. Hot yoga and meditation work for me. It may seem unrelated, but the first time you have something catching on fire in your system and you have to find it while mixing, or you have a small, inquisitive child loudly asking questions next to the console, or patrons get into a fistfight two rows down from you, your ability to keep paying attention to your show will save you. I see guys all the time that will be mixing and turn their heads at the late patron coming in, and then miss a pickup.

6. What Advice Do You Have for Working With Designers?

Designers are very loyal. As you find the people you work well with, cultivate those relationships. Always think what you can offer them to make their jobs easier, what you can bring to the table.

If you have a problem, like a speaker isn't going to fit where we hoped, or a piece of gear turns out to be incompatible for some reason, come up with a couple possible solutions to present along with the problem. Then don't get bent if they chose a different direction.

Be upbeat and positive. Tech is grueling for everyone, not just you.

7. What's the Best Note or Bit of Advice You Were Given?

"Never underestimate the value of a closed mouth." I'm pretty sure that was Brian Ronan.

Also, "Don't come to New York without a job." That was Steve Kennedy.

8. Do You Have a Good Mixing Horror Story?

Oh yeah. At the end of my preset for *Jersey Boys* one night, my CCM scrambled. I know nobody knows what that is anymore—it's the on-board brain module of a Cadac. So, I swapped the spare CCM in place and it scrambled too. Without the CCM, not only is there no automation, but the console doesn't know what it is. They were already holding the opening of the house for me, so I had to punt. I got on the phone with Gary Stocker and he told me how to remap the console manually. After that, if we were going to have a show, I was going to have to mix it manually. No channels would be reassigned anywhere—wouldn't change groups, which was tricky because we were an A/B system; wouldn't change auxes; wouldn't change VCAs. There would be no sound effects because the routers weren't functioning. And no inputs would switch, which meant when the band had their songs where they were wireless onstage, I'd have to switch those inputs manually too.

9. Do You Have a Good Mixing Success Story?

We pulled it off. The backstage guys helped me remap, which is tedious, and they told the performers exactly what was happening. They actually built a buzzer box to serve as the sound effects like door buzzers and phone rings and told me what actor they were going to stand next to so I could open a mic for it. And then I mixed the show that I knew so well, in a way that I'd never mixed it before.

The problem turned out to be a card in the main PC that went bad and was spewing trash downstream. Gary came in for the work call next morning and we found it after about an hour.

ANDREW KEISTER

Andrew is the kind of designer that I personally have so much respect for. He's done his time as an assistant and an associate and he has clock many hours behind a console as a mixer. I have never really had a chance to work with him, but I know him for his reputation of good solid work and great sounding designs.

1. What Are Some of the Shows You Have Designed?

Broadway: *Charlie and the Chocolate Factory, Southern Comfort, Tommy, Godspell, Company.*

2. How Did You Get Started as a Designer?

I studied theatrical sound design in college; back in those days there wasn't as much distinction between engineering

Figure 21.3 Andrew Keister.

and designing so I'm not sure I really comprehended what I eventually wanted to do as a career. But I loved mixing and I loved great sounding musicals, so I was game to do pretty much anything in that world. When I came to New York I started as an assistant designer—I really wanted to mix, but I didn't have a union card. Eventually I ended up mixing for a decade on Broadway before I moved fully into the design sphere, though I tell myself I'm going to go back to mixing someday (I do miss it!).

3. What Was Your Path to Designing on Broadway?

I vaguely remember that interview . . . and a few pints being consumed during it! The interesting thing for me in retrospect is that I was so focused on the goal of designing a Broadway show at a relatively young age, and I was so happy and proud to have achieved that milestone at 28 years old, but I was utterly unprepared for it. Not that the show didn't sound good—it was fine. But I was not fully formed as a collaborator and there was so much about the art and science that I didn't understand, not through lack of intelligence or training, but through lack of years at the grindstone. I immediately went back to mixing after that show because I recognized that there was so much more to learn from the process. So much more to understand by sitting amongst literally thousands of audiences and observing (and participating in) the interaction between performer and spectator. There's a spark of magic that happens with live performers in a room with a real audience that cannot be captured in film or on TV, and it's a fascinating and highly educational thing to observe in great detail.

4. What Do You Look for In a Mixer?

If you take a look at the 50 best mixers of Broadway musicals in the past 20 years (and that's an impossible list to define, but you get the idea), I think there are some remarkably similar personalities and traits. After a few years of trying to hire good mixers (with some notable misses in those years), I started looking for people who had similar traits to people who had been very successful at the job. So, it's not always important to me what a younger mixer has done as it is how well I feel their personality fits the job. Musical skills and talents are obviously very valuable, as are technical skills, but the ability to calm a high-strung performer [and] instill confidence in a demanding director is equally important.

5. What Do You Look for in a Deck Sound Person?

While I recognize there are very valuable deck sound people without aspirations to be mixers (and have some dear friends who fit that mold), I've always viewed that position as a training ground for younger mixers coming into the business. I toured as an A2 with a great teacher in the A1 job who really impacted my thinking about how that role can be used. In commercial theater in the United Stated there is sadly little financial incentive for producers to invest in the training of the next generation of mixers who will make their shows sound great. So to me, that is the one place I can put someone who I have a good feeling about eventually becoming a great mixer where they can earn a good living while learning the next set of skills I need them to have. Of course, the A2 job is a role in and of itself, and I don't want someone who is so keen to get out front that they aren't interested in doing the job they have exceptionally well, so it is a bit of a balancing act.

6. What Advice Would You Give to Someone Trying to Make It as a Mixer or Deck Sound?

I think observation is very important. See lots of shows. Participate in lots of theater. Study the art form. I never had any desire to be an actor, but I studied acting because I wanted to understand what the people I was working with were going through. I never wanted to be a singer, but I studied vocal performance because I wanted to understand how a singer did their job and what they needed to do it. You will be so much better equipped to be a collaborator if you truly understand the needs and processes of your partners in crime.

7. What Are Your Favorite Tips and Tricks for a Mixer?

I don't know that I have any great, magical tips. It's hard work to mix a Broadway musical. The keys to being successful are hard work, talent, and grace under pressure.

8. What Do You Expect From a Mixer?

I put a lot of trust in someone when I hire them to mix a show. We share success and we share failure. A lot of my job as a Sound Designer is to set my mixer up for success. I need a mixer who is capable of doing the job, who is willing to work long and hard, and who understands the process.

8. What Do You Expect From a Deck Sound Person?

I need someone backstage to be my eyes and ears to what's happening on deck. The better I can communicate with and understand the deck person, the better my understanding of the situation onstage. I need her or him to come ready to work, I need them to be dedicated to the job, and I need them to be eager to learn.

9. Do You Have a Good Mixer Horror Story?

I don't really have one—I've certainly had mixers who have not lived up to my hopes, but that failure is part mine as well. Sometimes you hire a great person, but it's not at the right time in their life. Sometimes you just hire the wrong person. Hire enough mixers, you're going to get some duds. Minimize the damage and move on.

10. Do You Have a Good Mixer Success Story?

No great stories really. I've had a few mixers I've hired on a hunch and it has worked out really well. I met one when he was a sophomore in college and we kept in touch and he interned for me, moved on to A2 jobs and eventually opened shows for me as a mixer. It's been a very rewarding process for both of us and I believe it will continue to be so for many years. I'm always on the lookout for the next great mixer.

STEVE KENNEDY

Steve is one of the legends in this industry in my opinion. He has designed so many huge hits and every show I have ever seen of his as sounded stellar. I met Steve for the first time when I was mixing *Jersey Boys*. I had been mixing the show for a month or so as the one show a week sub.

Typically, a designer gives the sub mixer time to get comfortable mixing a show before coming to note the mix. It is also not uncommon for actors to complain about the sub mixer mixing differently, even if the mix is absolutely identical. Steve showed up right before the downbeat and he asked if he could whisper notes to me while I mixed. I said of course he could. As the cue light for the start went off he leaned forward and whispered his favorite tip to me. Some things stick with you and having him whisper that to me is a highlight of my career.

1. How Did You Get Started as a Designer?

After engineering shows for some years, I just found it a natural progression. Working with great designers to learn from certainly helped. Being a production engineer and setting up shows on the road was also a huge learning experience.

2. What Do You Look for In a Mixer?

Calm, musical people with good ears and better instincts.

3. What Advice Would You Give to Someone Trying to Make It as a Mixer or Deck Sound?

Learn all you can about the equipment you will be using. Stay organized and know your show.

4. What Are Your Favorite Tips and Tricks for a Mixer?

Don't fuck up.

5. What Do You Expect From a Mixer?

Consistency.

WALLY FLORES

I knew of Wally since the day I moved to New York, but I didn't have the chance to work with him until about two years ago. He is a rock-solid mixer and Production Sound and he takes some great pictures. He took the picture that is the cover of this book.

Figure 21.4 Wally Flores.

1. What Are Some of the Shows You Have Mixed?

Broadway: *Little Shop of Horrors, Blood Brothers, The Rocky Horror Show, Macbeth, The Humans, Indecent, M Butterfly, Sylvia, Annie, Falsettos, Man of La Mancha.*

2. How Did You Get Started as a Mixer?

I went to school to study studio recording on a lark. I was hooked from that day on. I worked on a tour of *Man of La Mancha* that eventually came to Broadway. From then on, it was about learning my trade from the ground up.

3. What Advice Would You Give to a Young Sound Person Trying to Make It?

Open your eyes and your ears before you open your mouth.

4. Which Show Was Your Favorite to Mix and Why?

The Rocky Horror Show. The PA had serious horsepower, the desk was a Midas XL3 and that show was fun to mix.

5. What Are Your Favorite Tips and Tricks as a Mixer?

Be in the moment, even if your show is programmed to the nth degree.

6. What Advice Do You Have for Working with Designers?

Try to be as diplomatic as possible even if what they are suggesting doesn't make sense. And always bring your A game.

7. What's the Best Bit of Advice You Were Given?

"Stop trying to counter the big moments. Let them sing out."

8. Do You Have a Good Mixing Success Story?

"Are those actors really mic'd?" That came the other day. It was a compliment to the whole sound team as what we all were working towards was actually coming across to the audience.

PATRICK PUMMILL

Patrick has been a close friend of mine since college and since before I knew I was going to end up mixing musical theater. He is the kind of detailed perfectionist mixer that I was more like. His mixes are so perfect and well-thought out. It is a pleasure to watch him mix and he is the person I call when I am frustrated or elated in this business. He taught me the mix for *On the Town* last year. It was the first time we had worked together in that capacity. I don't know who was more nervous. I was terrified that I wouldn't be able to duplicate his mix and he was worried that I would see all the flaws in his mix, which didn't exist.

1. What Are Some of the Shows You Have Mixed?

Broadway: *On the Town, Thoroughly Modern Millie, Caroline, or Change, Follies.*

2. How Did You Get Started as a Mixer?

When I was in high school, I had access to a PA system that I loaned out to local bands. I discovered that I had a natural knack for it. I coupled that with spending summers as an apprentice at the local regional theatre, Casa Manana in Fort Worth, Tx., and I eventually ended up as a mixer for theatre. I spent multiple years working at a regional theatre and through a very lucky introduction to a top tier sound designer, Tony Meola, I got a temporary sub job on a national tour, which was *Les Misérables*. That led to more jobs and eventually shows on Broadway.

3. What Advice Would You Give to a Young Sound Person Trying to Make It?

Observe and listen much more than you speak. You'll learn something new every day. Never try to prove what you think you know to those in positions above you.

4. What Was Your Favorite Show to Mix and Why?

The NY City Center Encores production of *Most Happy Fella*. It was a semi-staged production that ran for just one week. The show used the original orchestration that called for an orchestra of 38 musicians. It was a remarkable sound.

5. What Are Your Favorite Tips and Tricks as a Mixer?

a   I prefer to be off-book as quickly as possible.
b   Consciously avoid muscle memory.
c   Any move that gets repeated every show can and should be automated, excluding fader movement.

6. What Advice Do You Have for Working With Designers?

a   Cultivating trust between a designer and mixer is of paramount importance. And it has to work both ways. The assessment of quality and appropriateness of sound is very subjective. Because of the collaborative nature of musical theater, many differing opinions can make creative decisions difficult. Honesty in conversation between mixer and designer helps to maintain that trust.
b   Work with designers whose aesthetic mirrors your own. If you don't like the way a particular designer's show sounds, then you will not be successful at mixing their shows.
c   The designer and the mixer should always be two separate people.

7. What's the Best Bit of Advice You Were Given?

You have two eyes, two ears, and one mouth. Look and listen twice as much as talk.

JORDAN PANKIN

Jordan taught me my first mix on Broadway, which was *Man of La Mancha*. I wish everyone had the chance to learn a show from him. Jordan makes everything look easy and he has such a joy for his job. Years later he taught me to mix *Bombay Dreams*, which was a decently complicated mix, but it was just as important to Jordan that I mix the show properly as it was that I did a hand wave to the cast at a certain moment of the show. It's a hard job, but it is also a fun job. I'll never forget something he said to me as I was learning *La Mancha*. He said, "My job is to make sure you succeed. I don't get to go on my vacation if you fail." It is a lesson I have carried with me to every job. My job is to help people to succeed and I don't succeed by others failing.

1. What Are Some of the Shows You Have Mixed?

Broadway: *Sound of Music, Kiss Me Kate, Steel Pier, A Funny Thing Happened on the Way to the Forum, Lion King, Bombay Dreams, Wicked, Nine, Jelly's Last Jam, Sweet Smell of Success, Moon Over Buffalo, Man of La Mancha, Les Miz, Busker Alley, Grand Hotel, Sophisticated Ladies, Annie Warbucks.*

2. How Did You Get Started as a Mixer?

It was the end of the school year at SUNY Binghamton; my friend who usually mixed the local summer stock season got a better job offer on Long Island and he asked if I'd like the sound mixing job. I said sure and I spent the summer mixing 3 musicals and 2 plays.

3. What Was Your Path to Mixing on Broadway?

Figure 21.5 Jordan Pankin.

I ended up working at BAM (the Brooklyn Academy of Music) and got other job offers from there for small tours and eventually a summer stock gig at Candlewood Playhouse in Connecticut. It's there that I met Lew Mead, Bob Rendon and afterwards worked for them at Promix where I met Tony Meola, Otts Munderloh and other great designers. Eventually, I was offered a national tour which led to an offer for another tour which was supposed to come back to Broadway. This fell through but another musical popped up quickly and I was mixing on Broadway.

4. What Advice Would You Give to a Young Sound Person Trying to Make It?

Keep your eyes open, hands out of your pockets and show up on time. Watching what the experienced mixers and designers are doing can really help the new people see what to do in so many situations.

5. Which Show Was Your Favorite to Mix and Why?

Kiss Me Kate, sound design was clean and clear and crisp and natural and the cast was incredible.

6. What Are Your Favorite Tips and Tricks as a Mixer?

Listen, layer your mix, breathe with the actors, understand what they are trying to do, befriend all departments, dynamics, keep your console clean and keep your script cleaner.

7. What Advice Do You Have for Working With Designers?

Protect your designer, take their notes and see what works; never play a new sound cue without the designer present, be part of the team, not the competition.

8. What's the Best Bit of Advice You Were Given?

Make it sound good where you are mixing from; listen; learn who everyone is on a production and what their story is and where they are in the pecking order. You never know who you're talking to.

9. Do You Have a Good Mixing Horror Story?

I was mixing a big show in NYC and the backstage sound man had messed up the same over and over again and I had had enough. I pounded the console and that impact unmuted an input channel with an unsquelched wireless on it. It was really bad, the audience chanted they couldn't hear and we had to stop the show to power up again. 5 minutes. Totally humiliating, but I never hit a mixing console ever again.

10. Do You Have a Good Mixing Success Story?

Yes, anytime I finished the show without missing a pickup, when all the elements fell into place, vocally, instrumentally and technologically and there was no equipment failure, no loud candy wrappers, no loud ice in plastic cups. The orchestra wasn't too full of subs and the cast was well rested and watching the conductor closely. Basically, the stars aligned and everything came together to make my job a bit easier.

BRIAN SHOEMAKER

I would easily consider Brian one of my closest friends in this crazy business. I hired him for a tour of *The Full Monty*. When I decided to hire him, I was not sure if it was a good idea or not. I called to interview him and he told me he had just finished a tour as the system tech for Motorhead. Now, I love me some Lemmy just as much as the next person, but I wasn't sure if a system tech for a hardcore rock band was the right choice for musical theater. I finally decided to hire him because his resume was mostly musical theater and we got along really well on the phone. He turned out to be a much better mixer than I had ever expected and I put him in the top three or four mixers I have ever watched mix a show.

1. What Are Some of the Shows You Have Mixed?

Full Monty, My Fair Lady, Avenue Q, Spamalot, Spring Awakening, Next to Normal, Book of Mormon, If/Then, Jersey Boys, and *Mean Girls*.

Figure 21.6 Brian Shoemaker.

2. How Did You Get Started as a Mixer?

I used to go to a lot of concerts when I was in my teens. I always thought it'd be cool to be the guy in the middle of it all, traveling the country and getting paid to listen to great music.

I didn't know how to do that, so I went to college for something else completely different. One year in, I discovered technical theatre and a path to mixing sound and here I am.

I went to Wright State U (Dayton, OH) and received a BFA. While there, I worked at a lighting/sound company based in Cincinnati. I got touring experience through that job and mixing experience in theatre. In 2003 I got a call from Shannon Slaton about a job to mix *Full Monty* Tour. Several tours, a house job in Las Vegas and being married to a Broadway actress eventually brought me to NYC to mix on Broadway.

4. What Advice Would You Give to a Young Sound Person Trying to Make It?

Work. Work as much as you can in your field. Don't take any other job but sound positions. Be personable and treat people well. It's a small world.

5. Which Show Was Your Favorite to Mix and Why?

Jersey Boys, because it was an incredibly challenging musical, but incredibly rewarding if done correctly. Plus, you got to hear amazing music every single night!

6. What Are Your Favorite Tips and Tricks as a Mixer?

Understand the physics of sound. And realize you're better off being friends rather than enemies.

7. What Advice Do You Have for Working With Designers?

When it all comes down to it, it's their name in the program. As a mixer, it's your job to maintain the sound design. Don't take your mix personally; it's a collaboration. Be accepting of notes and work towards perfecting the mix.

8. What's the Best Bit of Advice You Were Given?

Always say Please and Thank You. Remember people's names. And try to work harder than the person next to you. All from my dad.

9. Do You Have a Good Mixing Horror Story?

I've mixed the classic 'Open Mic in the bathroom' show. We had the original PM1D programming from Broadway on tour. I was on the B mic of our main actor during the show.

I was mixing a scene about suicide and I heard a steady stream through the system. When the music stopped, it was obvious what it was. I flipped to the other page and sure enough, Main Actor was wide open with no VCA in line. Unity to the system of a good long pee. Thank god it didn't turn out like *Naked Gun*. To top it off, I was recording.

10. Do You Have a Good Mixing Success Story?

Straight out of college, I was mixing a show at Williamstown Theatre Fest. Everyone knows how crazy that turnaround is. As I was at the console, I was thinking to myself, did I make the right decision by doing sound in college. Should I have done props or something else. Then a call came out over radio that the porto-potty that was set up in the DSL wing for our legendary (but elderly) actress accidentally tipped over and props had to clean it up. I knew then, I made the right choice by being a sound guy.

Scott Anderson

There are certain people you meet in life who make you want to be a better person. There are other people who make you want to be better in your career. And then there are people who make you want to be better at both. Scott is one of those rare people. I can easily say that meeting and working with Scott was one of the most rewarding experiences in my career. When I mixed *Annie* at the Palace, he was the deck sound and sub mixer. Scott had already been mixing and working on Broadway for about 30 years when we met. Training him to mix a show was the easiest experience I have ever had in training a new mixer. It was effortless to him. Scott has a passion and a true love for mixing musicals that seems to radiate from him. And that passion seems dwarfed by his devotion to his family. This is a tough business. It is hard to have a successful career and a successful marriage because of the odd hours and insane work schedules in this business, but it is possible. I learned a lot from working with Scott, but the most important lesson he taught me was in something he said to me that I have repeated thousands of times. "The one thing you'll never regret is taking time off to be with your family." Nothing could be truer.

1. **What are Some of the Shows You Have Mixed?**

   The Mystery of Edwin Drood, Me and My Girl, Miss Saigon, Beauty and the Beast.

2. **How Did You Get Started as a Mixer?**

   I was working as a stage manager for an amusement park and forced to run the sound. I loved it and never left.

3. **What Was Your Path to Mixing on Broadway?**

   I toured on both bus and trucks and 1st national tours before being asked to open a show on Broadway.

4. What Advice Would You Give to a Young Sound Person Trying to Make It?

Work hard in the shops or on tours to learn the equipment and ask for every opportunity to learn to mix. If you have the talent for mixing you will be discovered.

5. Which Show Was Your Favorite to Mix and Why?

Les Misérables, has a terrific score with sweeping dynamics and wonderful orchestral moments that are pure fun to work with.

6. What Are Your Favorite Tips and Tricks as a Mixer?

Strive to develop a consistent vocal mix that supports the dynamics of the performers; this allows you to use the orchestral dynamics and the orchestrations to interpret the expressive quality of the music.

Figure 21.7 Scott Anderson.

7. What Advice Do You Have for Working With Designers?

All designers are different, some want you to be a partner, others just expect you to do as your told. Keep your line of communication open and honest so you can build the trust needed to succeed.

8. What's the Best Note or Bit of Advice You Were Given?

From a director: "You have to anticipate the ad-libs," from a co-worker: "Stop missing pick-ups," from a designer: "your mix needs to match the action on stage."

9. Do You Have a Good Mixing Horror Story?

There was an old "feature" on early CADACs called ALL BYPASS which removed all faders from VCA control. During a power glitch the desk went into that mode. This meant that all faders were on and at 0; the resulting feedback was extraordinary.

10. Do You Have a Good Mixing Success Story?

After a performance of *Les Misérables* at the Imperial Theater an audience member stopped by to say he had not heard a show sound this good since he saw *Chess* in this theater. I had been the mixer for both.

22

SUGGESTED PROJECTS FOR MIXERS

I have worked with a lot of mixers just starting out and there are two projects that I wish they had all worked through. I have intentionally avoided drifting too far into the world of design because that is not the purpose of this book, and I have intentionally avoided going into depth on technical issues such as compressor settings or EQ because that is also not what this book is for. But I am going to dip my toe a little to explain these projects because I think it is important that a mixer have a well-rounded understanding of the sound world around them.

The first project is to design a sound system on paper for a theater you've never been in. It is very common for me to work with someone right out of college who has never taken the time to think through a sound system design outside of their college theater or regional theater. It would be very helpful to spread your wings and force yourself to think of a system bigger than you have ever used. So how should you go about this?

First things first. Find a theater to design a sound system for. If you do a search for "Broadway Theater Floorplans" you will find a treasure trove of theater floorplans. Pick a floorplan and get started designing a sound system. If you can find a floorplan and a section of the same theater it is even better. The first task with this project is to draft the theater in some piece of software. Vectorworks is definitely the industry standard at this time. I believe the educational version is free, so the buy-in for this project is pretty low. If you are studying sound design, you should have a basic knowledge of a drafting program. I worked with some interns recently who were about to graduate college and had never used a drafting program. That's crazy. Basic drafting is essential. The first thing to know about Vectorworks is that it is considered a 3D modeling program. You can draft in 2D and 3D, but if you want to be a rock star then learn to draft in 3D and take the floorplan and section drawing and create a 3D model of the theater.

I have drafted almost all of the Broadway theaters into Vectorworks in 3D. It probably takes me about eight hours to draft a theater in 3D. That's not that much time and the payoff is huge. I can add my speakers in 3D and spin the theater around to see sightline problems. I can turn the speakers on and see what the coverage will look like. Once I have my drawing in 3D, I can create 2D viewports of the theater. But if you don't want to take the plunge into 3D drafting, then at least draft the floorplan in 2D.

Okay, now that you have a drawing to work with, it is time to add some speakers. It is time to plot out how you would design a sound system for a musical in the theater. The most basic objective for any musical theater sound designer I know is to make sure every word is heard. Most designers strive for even coverage. There are other roads you can go, but my advice is to go with the basics. Design a system with even coverage. And the best part: You have an unlimited budget and no scenic or lighting designer to tell you that you can't hang a speaker there.

One basic question is whether you should use powered or unpowered speakers. I suggest going with unpowered, so you can learn more about amplifiers. Put your speakers on the drawing and draw out the speaker patterns. What is your goal SPL? Is it 80dB, 85dB, 90dB? Make a decision and then confirm you can achieve that to every seat in the venue. How do you do that? Well, that is why I am suggesting this project. This book is not intended to delve to deeply into topics that require several books to cover. But that is the point of this project. Every sound person should be able to do this. If you aren't, then pick up some books on system design. It will make you an infinitely better mixer.

There are some basics you can use to plot out a sound system. First is the Inverse Square law. You lose half the volume when you double the distance. If you know a speaker is 100dB at 1 Watt at 1 meter then you can predict what the volume of that speaker will be at 16 meters from the speaker. Since you double four times and 6dB is double or half the volume then you would lose 24db at 16 meters. That would make the speaker 76dB at 1 Watt at 16 meters. So how many Watts of amplification do you need to get that speaker to our desired SPL of 85dB? The equation for this, which can be found on Crown Amplifiers' website, is:

Equations used to calculate the data:

$$dBW = Lreq - Lsens + 20 * Log(D2/Dref) + HR$$
$$W = 10 \text{ to the Power of } (dBW/10)$$

Where:
Lreq = required SPL at listener
Lsens = loudspeaker sensitivity (1W/1M)
D2 = loudspeaker-to-listener distance
Dref = reference distance
HR = desired amplifier headroom
dBW = ratio of power referenced to 1 Watt
W = power required

You can find a calculator for this at www.crownaudio. com/en/tools/calculators. I'm not going to tell you the answer. You can figure it out.

But what if you don't want to turn that speaker up? I mean, 16 meters is a long throw in a theater. Maybe it would make more sense to add a delay speaker to make up the difference in SPL. By a delay speaker I mean a smaller speaker for under the balcony or a fill speaker in a corner. It is typically called a delay speaker because we have to add delay to line up with a speaker at the proscenium. The idea being that we delay the sound wave from leaving the underbalc delay speaker until the sound wave from the proscenium speaker has time to reach that speaker. By doing that we can line up the sound waves so that they work together instead of fight each other.

This is probably the most important thing you can learn about theatrical sound system design. You can guess it using math on paper, but it is best to test for it with a fully installed system because every component in the system could add different amounts of latency, or delay. Some people use a laser disto in the theater to figure out delay times. Again, that is just a rough guess. You can only figure out time delay by measuring the actual equipment.

To do a rough guess about time delay between two speakers, you can do a simple division problem. Measure the distance between the two speakers. Let's say it is 14 meters. The speed of sound is 343.2 meters per second. Simply divide the two and you have a rough guess at what amount of delay should be put on the delay speaker. But you haven't considered the latency of the rest of the equipment or the altitude or humidity of the theater. So this is a rough guess, but good enough for predictions on paper. If you use a laser disto in a theater, you are ignoring just as many variables and are merely dealing with the difference of distance between the speakers and not the actual latency between the two speakers.

There are ways to properly test for delay. You can use one of many pieces of software built for system tuning. These include Smaart, Simm, and Tuning Capture. This is the best way to measure delay. If you don't have that software, there is a trick you can use to find the delay time. It is not as accurate as the software, but it is more accurate than simple measuring. Use a metronome that is patched to two inputs on the console. Flip polarity on one input. Send the normal metronome to the proscenium speaker and the flipped metronome to the delayed speaker. Turn the level up so that the two speakers are equal SPL at the delay speaker position. You will hear two distinct metronome clicks. Add delay slowly. When the signals line up, the polarity reversed click will cancel out the normal click and the metronome will get much quieter. If you add too much delay, the click will start to get louder again until it is back to normal and two distinct clicks. Find the spot where the clicks are one and the quietest and you found the delay time.

Continue adding speakers and figuring out SPL and amplification needs and delay times. The other aspect you can predict is the interaction of speakers. If you have a proscenium left and right speaker, then the two speaker patterns will overlap somewhere in the center of the theater. As we move off axis of the speaker, or away from the center, we lose volume. Each speaker is different, and you can find this information in the datasheets for the speakers you are using. As an example, let's say our left and right speakers lose 6db at the point where the two speakers begin to intersect at the center of the theater. Let's say that the left and right are 85dB on axis an 79dB with the reduction off axis. When two speakers interact, there is a simple calculation that can be found in table form for what the SPL would be combined. In this case, the area where these speakers meet would sum together and add 3dB, making that area 82dB in the center of the theater. We could use the proscenium center speaker to add 3 dB to that area to level it out so that each area is 85dB. Of course, to do that we have to get a little more advanced and look at the summation of three sources. In this case, the center would need to be 82dB and with the three interacting you would get an SPL of 85dB. You can find more information online at websites like this: www.sengpie laudio.com/calculator-spl.htm.

Difference between the two levels to be added in dB										
0	1	2	3	4	5	6	7	8	9	10
3.01	2.54	2.12	1.76	1.46	1.19	0.97	0.79	0.64	0.51	0.41

I want to make sure that you understand that these are just some basics. I strongly encourage you to read some books that really explain this in more detail, but for our purposes, this is sufficient. So now that you have laid out the system and have even coverage all over the theater you are done, right? Heck no. Now the fun begins. How many amps do you need? What are the names of the amp channels? What cables and lengths do you need to plug this system together? What kind of speaker yokes do you need? This one project could go on and on, and I wish everyone I met with a degree in sound could open their portfolio and show me a system design.

Okay, so now that you've figured out the cable and speakers, it is time to move on to the important stuff. Intercom and video. We will skip consoles and playback and mics. That's just a simple list. But intercom and video. Now that is something you need to understand. So I suggest you lay out a video system. Where do you need cameras? Who needs to see what? What kind of cameras? What lenses do you need? I will be honest. I hate dealing with video. And oddly enough, I am the guy who revolutionized the way we distribute video for theater on Broadway and tours. I don't know if there is a show out there, currently, that isn't using an ETS SDS887, which is a four-camera video distribution amplifier to eight locations over Cat-5. In case you are curious about the model number, my initials are SDS. But I still hate video. It is a necessary evil for us sound people. Go read up on cameras. There are plenty of online calculators to figure out lens size for distance. The basics for a musical are: a conductor camera, a front of house color camera, and a front of house low-light black and white camera. We love and hate HD video because it adds latency and can cause the cast to sing a beat late or cause the stage manager to call a cue late. I would suggest sticking to the analog world.

And then there is com. My trusty old friend. Always there to make my day worse. My final suggestion for this project is to plot out an intercom system. You will need public channels and private channels. You will need to read up Clear-Com main stations to see how many remote stations and belt packs a main station can power. You will need to figure out who gets what channel and how to cable it all. The normal channels for a basic musical are: Deck, Lights, Spots, Sound, Lights Private, Moving Lights Private, Spots Private, and Sound Private.

Now that you have thought through the design of a sound system, it is time for my second project suggestion. This one should be obvious. Learn to mix a musical. That seems appropriate, doesn't it. I think this project should be done over and over again. This is what I mean. You don't need to mix a physical production of a musical to learn to mix a musical. Some people think you need a multi-track recording of a musical in order to learn how to mix a musical. Wrong. All you need is a script and a Broadway cast album. I have learned dozens and dozens of shows from other people. Whenever I learn a show I am given a script and a recording. I watch the mixer mix and document their mix in the script, but you will be creating your own mix. When I am learning a show, I practice by listening to the recording and pretending to mix. If anyone really wants to learn how to mix then you should start doing this now. Mixing is so similar to playing an instrument. The more you practice moving the faders up and down and learning where to place your hands, the better you will be.

There are plenty of great musicals out there to practice with. You could choose to mix songs off of a Broadway cast album or you could look for videos of full Broadway productions. There is actually a great resource for this right now, which is BroadwayHD.com. This website records and streams Broadway shows. I mean how incredible is it that there are full recordings of Broadway musicals at your disposal? So what are you waiting for? Pick a show and get to mixing. Some of these shows are classics and very easy to find a script for. If I could suggest a show for you to start with it would be *She Loves Me*, which was done at Studio 54 by Roundabout Theater Company. I had no part of this production, but I did design the sound system that is used for most shows at Studio 54. *She Loves Me* is one of my favorite musicals, and it has a scene at the beginning that is perfect for learning to mix. It is a scene with three shop attendants and three customers. It is a workout for your fingers.

Whenever I learn a show, my first step is getting the script right. I always try to get a script in electronic format, whether it is Word or PDF. Then I can edit the script to be exactly what I want. I have also typed several scripts myself whenever I couldn't get an electronic version. The script for *She Loves Me* can be purchased in printed form, but with a little extra effort you can probably find it in PDF form. Either way, I would recommend getting it on your computer in an editable form. I would then listen to the recording of the show and update the script for every line change and every unscripted laugh or sigh. In other parts of this book I have explained what a script should look like and how to mark up a script for line-by-line mixing. That would be the next step.

Once you have a script, then put on some headphones and see if you can keep up with the show. When I learn a show, I usually take about two weeks. I practice one scene over and over again until I can nail it effortlessly. I then add the next scene. It's great to have the console you are going to actually mix on to practice with, but it isn't necessary. I have learned to mix a show using quarters on a table as faders and a dime for the "Go" button. I have used a Mackie 1604 a lot as a practice board. Figure 22.1 shows me with my normal setup for practicing. This was in 2007 and I was learning *Legally Blonde.*

When you are practicing mixing, notice when the band swells. Notice what happens during scene changes. Document sound effects and add them to your mix. Mixing is all about consistency. Since you don't know how the board was set up then, you should just decide on how many VCAs you have. Normally it is 12. You should also decide where the fader will be for dialogue and where it will be for singing. With the orchestra, you can decide whether you want a single "Band" fader or to break the band up into multiple VCAs. Then decide where the band fader(s) should be for underscore and for songs and for dance breaks.

Once you feel like you have mastered it, have someone put on headphones with you and let them watch you mix and note your mix. Maybe they will see how you are tripping yourself up because of programming or have a suggestion on how to make the mix easier.

The bottom line is that if you want to be a mixer, then you should do this project.

Figure 22.1 Me and my son Parker, who wanted to learn how to mix a Broadway show.

EPILOGUE

If you have made it this far, the only thing left to talk about is running the show and loading it out. The challenge to running a show is management. You have to manage your time and plan battery orders and workcalls. You also need to maintain a good relationship with the shop to take care of gear as it breaks. Hopefully your show will run for years and if it does, you will have to deal with monotony and complacency. You will have to find a way to keep the show fresh and not drift from the designer's vision. Occasionally the designer will stop by and watch your show and note it, but the designer really hopes not to hear from you for a long time. The designer just wants a smooth-running show with no problems. And that is a challenge as the actors get bored and start nitpicking one department after another. Eventually all involved get their turn in the barrel. You also have to find and train a sub, which is an even bigger challenge. When I learned my first Broadway mix, the mixer told me, "I need you to be good at this because I am going on vacation with my family."

After your show runs for 20 years or a few months, it will be time to take it all down and send it back to the shop and start over on another show. We work in a business where our jobs are transient and we have to keep that in mind as we sit on our little musical. It is important to keep your name out there even when you have a job. And when you see the signs of the ship taking on water, you better call your designer and find out if there is another show out there. Classic signs are when the show cuts the workcalls from every week to every other week or starts printing the playbill cover in black and white instead of color. Luckily, the Broadway Grosses are posted online every week to keep stagehands on edge predicting when their show will close. Will it make it to January or will it close in September? Or the worst option . . . will it close the week after the Tonys, leaving you out of work for the very, very dry summer?

I hope you found this book an interesting read, and I hope it gave you a glimpse into the world of mixing musicals on Broadway. I hope you learned some things that will help you wherever you work, and if you come to New York to work I hope this book gives you a little leg up. It is definitely a different world here. There is a good career to be made in sound, but there is a learning curve to understanding what we do on Broadway. I hope this helps you over the hump. Of course, I am sure the first person you meet in New York will tell you everything in this book is wrong and he will be right. Sound is the most changing and unstandardized part of theater. Every shop has slightly different terminology and every mixer has his method. This book is glimpses of what I have found to be standard and what I have been taught by lots of people. There is no right and wrong in sound as long as it sounds good, and of course that all depends on the ear of the listener.

Finally, I want to share some wisdom from my good friend, Jordan Pankin, who mixes *Wicked* and is an excellent mixer. Whenever the stage manager called "Places," he would look at me and say, "Remember . . . Places starts with a pee."

INDEX

Note: Page numbers in *italic* indicate a figure on the corresponding page.

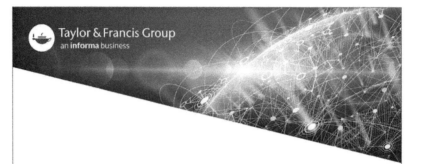

Milton Keynes UK
Ingram Content Group UK Ltd.
UKHW031950151223
434399UK00010B/44